ANTI-GRAVITY & THE UNIFIED FIELD

Adventures Unlimited Press

The Lost Science Series:

- THE ANTI-GRAVITY HANDBOOK
- ANTI-GRAVITY & THE WORLD GRID
- ANTI-GRAVITY & THE UNIFIED FIELD
- THE FREE ENERGY DEVICE HANDBOOK
- THE TIME TRAVEL HANDBOOK
- THE FANTASTIC INVENTIONS OF NIKOLA TESLA
- UFOS & ANTI-GRAVITY: PIECE FOR A JIGSAW
- THE COSMIC MATRIX
- THE A.T. FACTOR
- MAN-MADE UFOS: 1944-1994
- THE BRIDGE TO INFINITY
- THE ENERGY GRID
- THE HARMONIC CONQUEST OF SPACE
- ETHER TECHNOLOGY
- THE TESLA PAPERS
- QUEST FOR ZERO POINT ENERGY

ANTI-GRAVITY & THE UNIFIED FIELD

This book is dedicated to all scientist-philosophers in the world, where ever they are. It is my hope that these people will come together for the betterment and brotherhood of all mankind.

ANTI-GRAVITY & THE UNIFIED FIELD

Printed in the United States of America

Eighth Printing, September 2001

ISBN 0-932813-10-0 8/03

Published by
Adventures Unlimited Press
One Adventure Place, Kempton, IL 60946 USA

www.wexclub.com/aup
www.adventuresunlimitedpress.com

Cover painting © by Brooker
Courtesy of Mark Balfour and
Simply Living magazine, Australia

The illustration on the top of page 48 is courtesy of the Unarius Educational Foundation, El Cajon, California.
The illustration on page 35 is courtesy of The Kinetic Energy Foundation for Health and Healing, 1006 W. 10th Ave., Northglen, CO 80234

20 19 18 17 16 15 14 13 12 11 10 9 8

TABLE OF CONTENTS

Thanks to Mark Balfour, Bill Clendenon, John Walker, Linda & Richard Drane, Winston Whitaker, Moray B. King, Ken Beherendt, the Unarius Foundation, Carole Gerardo, Harry Osoff, Ted Martin, Hans Lauritzen, Brooker, Mark Balfour, the Tesla Society in Colorado Springs, Bob Powers, Al Freeman, the Stelle Community, and everyone else who helped with this book.

ANTI-GRAVITY
&
THE UNIFIED FIELD

A human being is a part of the whole called by us "Universe," a part limited in time and space. He experiences himself, his thoughts and feelings as something separated from the rest, a kind of optical delusion of his consciousness, This delusion is a kind of prison for us, restricting us to our personal desires and to affection for a few persons nearest to us. Our task must be to free ourselves from this prison by widening our circle of compassion to embrace all living creatures and the whole nature in its beauty.

—*Albert Einstein*

ALBERT EINSTIEN & THE UNIFIED FIELD

by

David Hatcher Childress

Lotte Jacobi

DR. ALBERT EINSTEIN

ALBERT EINSTEIN & THE UNIFIED FIELD

> *It would of course be a great step forward if we succeeded in combining the gravitational field and the electro-magnetic field into a single structure. Only so could the era in theoretical physics inaugurated by Faraday and Clark Maxwell be brought to a satisfactory close.*
> —Albert Einstein, *Mein Weltbild*

Recognized in his own lifetime as one of the most creative and influential minds in human history, Albert Einstein is perhaps the best-known scientist in the world, more than thirty-five years after his death. He is best known for his various Theories of Relativity in which he reexamined some of the most fundamental ideas in science and created a completely new outlook on the nature of space, energy, matter and time. While his Theories of Relativity are what he is most famous for, it is his work on the atomic bomb and the "Unified Field" that are Einstein's most important contributions of science. It is this last subject, being formulated for the public from the 20's up until his death in 1955, that has the most profound ramifications for science and mankind as a whole. It is the incredible theory, practice and implementation of the *Unified Field.*

Einstein was born in 1879 in Ulm, Germany to Jewish parents and was a poor student. Probably dyslexic, young Albert had a difficult time completing his assignments correctly, and was thought retarded by his teachers. Nevertheless, he learned to play the violin in his spare time, demonstrating a fine talent and developing a deep love of music that remained with him throughout his life. Indeed, it is probable that his love and knowledge of music, when combined with his mathematical skills, made him the genius he was.

In the 1890's his family left Germany for Switzerland where Albert completed his education. Except in mathematics (in which he earned his Ph.D.), his school record was poor, and it was not possible for him to find an academic post as he desired, so he eventually settled for a job as a junior clerk in the patents office in Bern. The work was so undemanding, Einstein found, that he had plenty of time for his own research. A pen and paper were his only equipment, and his fertile and probing mind, his only laboratory. Working entirely on his own, he formulated the beginning of a theory that was to shake the very foundation of science.

He started by looking again into the Michelson-Morley experiment on the speed of light in the "ether" and its strangely negative results. Albert Michelson (1852-1931) was a physics professor at the University of Chicago who designed the modern interferomter, with which he accurately measured the speed of light. He is also known for the Michelson-Morley experiment in the late 1800s which was to test the speed of light through what the then called the "Ether." The Ether was the theoretical fabric of the universe, and its existence by slowing down the speed of light.

The Michelson-Morley experiment was to prove to the scientific community at the turn of the century that there was no "Ether." Einstein was confused why there should the negative results of the Michelson-Morley experiement and his solution to the problem led him to think of relativistic physics. Curiously, the whole concept of "Ether" was to be revived at the later portion of the this century and called the "Quantum Field." Today, most physicists have accepted the theory of the Quantum Field, yet they also believe that the Michelson-Morley experiment disproved the Ether. It is a modern-day scientific paradox that escapes it's victims—virtually all modern scientists.

By simply manipulating ideas and following where the mathematics led him, Einstein produced a remarkable new picture of the universe. Published in 1905, the *Special Theory of Relativity,* as Einstein called it, challenged the view of time and space that had been accepted since the time of Isaac Newton. For over two centuries scientists had unquestionably believed that the basic quantities of measurement—mass, length, and time—were absolute and unvarying. Einstein showed that in fact they depended very much on the relative motion between the observer and what he was observing.

In 1915, Einstein then published his general theory of relativity which gave a mathematical description of space. He maintained that the universe consisted of a continuum of space and time in the form of a complicated four-dimensional curve. The implication of this difficult idea was that the force of gravity, first identified by Newton, was actually created by localized bending in the fabric of space, caused by the presence of large accumulations of mass such as stars and planets.

One prediction of relativity theory is that moving objects should show increase in mass, shrinkage in length and slowing down of relative time. At the speed of light any object would have infinite mass and zero length for time would stand still—predictions that Einstein believed were confirmed in the study of high speed relativistic particles.

A violent controversy was created by this new theory. Most scientists found Einstein's work incomprehensible, and even those who could follow the mathematics were unable to accept conclusions that seemed so contrary to common sense. But although Einstein had conceived the theory entirely in his own mind, he knew that certain experiments could help to prove him right. If publication of his ideas created a controversy, then the "proof" of his theories, published in 1919, caused a sensation! As scientists began to take his work seriously, the full measure of his achievement became clear. The young physicist had caused the greatest revolution in scientific thinking since that of Isaac Newton, formulator of the Universal Law of Gravitation.

Einstein was not entirely comfortable with the international publicity directed at him, but, like it or not, he had become a world-wide celebrity. The public regarded him as a man of unparalleled genius, and his name quickly became a synonym for great intellectual ability.

In 1914, Einstein had accepted a position at the University of Berlin as a Professor of Physics. He remained in that post for about 20 years, during which time he traveled widely in the Europe and the United States. He was a popular lecturer, speaking not only on his work, but also on social and political themes. Although he disliked public appearances, he used his name and fame to fight the rise of Nazism in Germany. He also advocated a the establishment of Palestine as a homeland for the Jews. He also backed the pacifist movement and other humanitarian causes.

When the Nazis finally came to power in 1933, his property and citizenship were taken away while he was on a trip abroad. Rejected by his homeland, Einstein was warmly welcomed in the United States. That same year he joined the

Einstein and Rabindranath Tagore

Michelson, Einstein, and Millikan

The spiral nebula in Coma Berenices, a distant island universe seen on edge. Its similarity to flying saucer shapes is significant, showing that the forces and that the vortex mechanics of the universe are operational from the smallest scale to the largest. Photo courtesy of Mount Wilson Observatory.

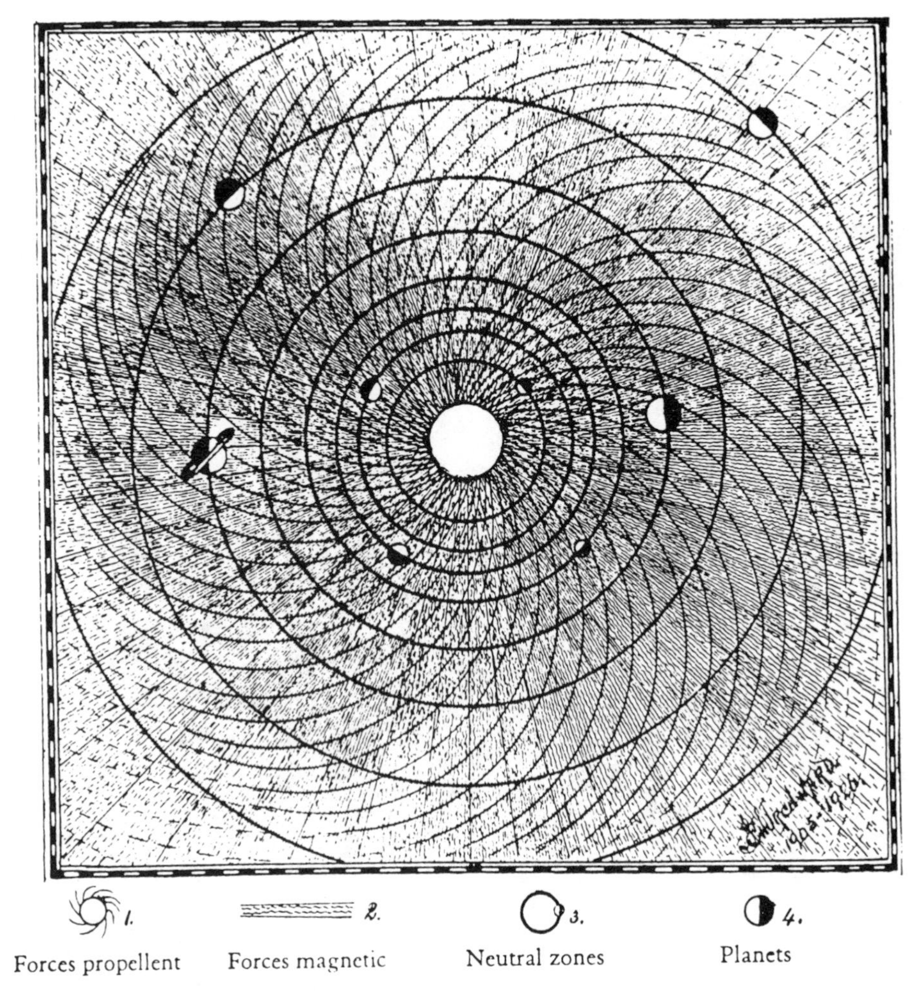

THE SOLAR SYSTEM

This is a conventional diagram. Distances and sizes are not considered.

1. Curved lines radiating from the sun. The centrifugal force.
2. Straight and wavey lines. The sun's forces, including magnetic.
3. Black circles. The neutral zones of the planets.
4. The planets.

From *The Cosmic Forces of Mu* by James Churchward, 1934. Churchward forsaw the now accepted belief that all the planets swim in a vast magneto-gravitic field around the sun.

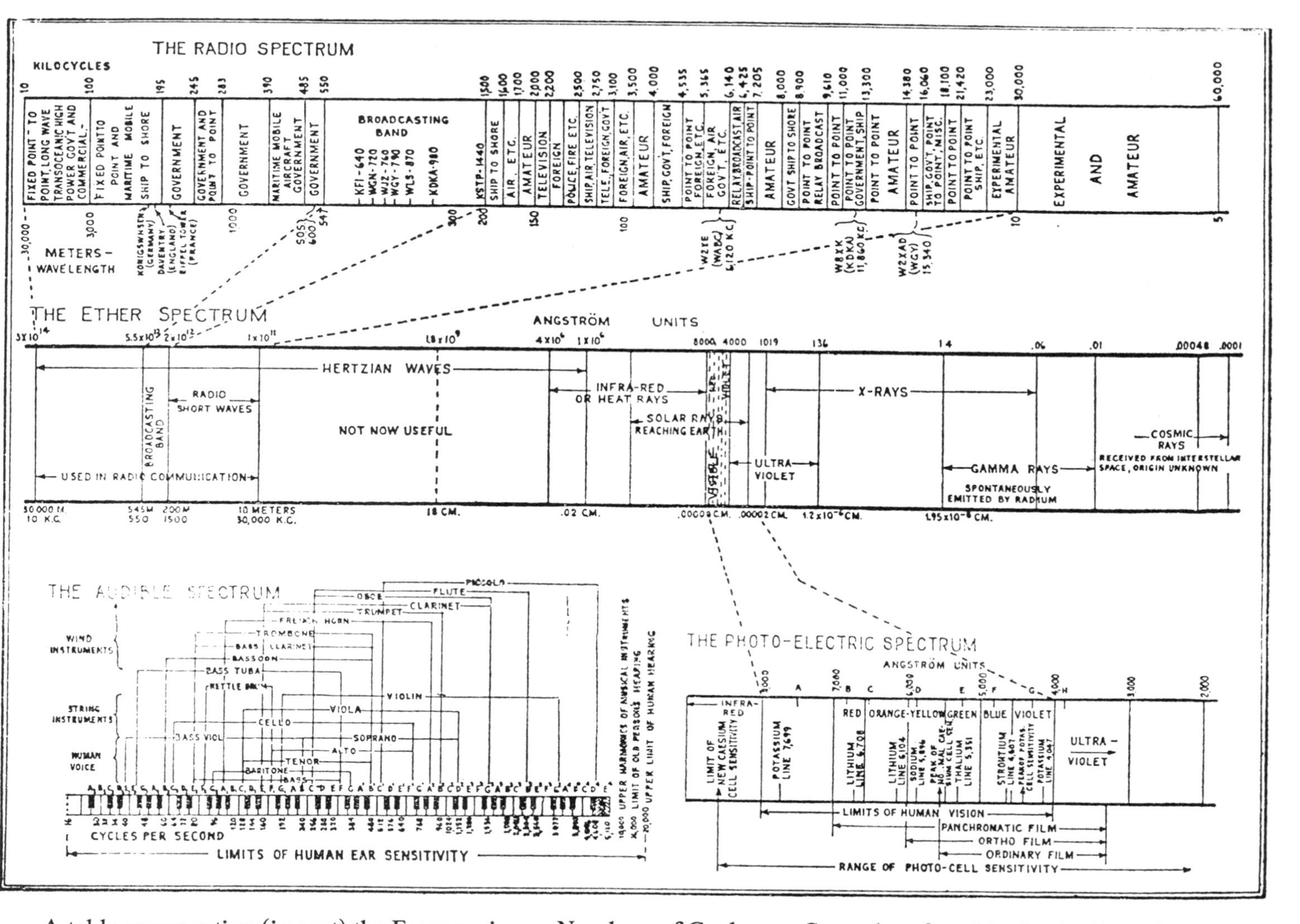

A table representing (in part) the Frequencies — Numbers of Cycles per Second — found in the Audio and Electromagnetic Spectrum. All physical matter and effect is a wave form of one kind or another, including Gravity. From *MAGNETISM and Its Effects On the Living System.*

Institute for Advanced Studies in Princeton, New Jersey, and remained there for the rest of his life.

Hitler himself attacked Einstein after Einstein had settled in the United States, saying that no Jew could formulate the Theory of Relativity. Hitler even suggested that Einstein had stolen the concept from papers carried by a German army officer who had been killed in World War I. By 1939 American scientists were becoming aware of the dangers of Nazi Germany and its aims, and that the Theory of Relativity could be applied by German scientists to build a devastating new weapon. They based this fear on the aspect of the theory showing that mass could be converted directly into energy, and that a minute piece of mass could release a vast amount of destructive energy. This opened the possibility of an immensely powerful new kind of bomb.

Under the threat of another world war, American scientists persuaded Einstein to write to President Roosevelt to suggest that the United States develop a counterweapon. Torn between his pacifist beliefs and his deep opposition to Nazi oppression, Einstein agreed—partly because he never expected such weapons to be used except as a deterrent. His letter led directly to the building of the first atomic bombs and to their use against Japan in 1945, despite Einstein's desperate last-minute appeal that such a devastating weapon should not be dropped.

Einstein spent his last years in semi-retirement in Princeton where he continued to teach and work on his most important, and largely neglected or suppressed theory, the Unified Field Theory.

Einstein's Unified Theory

I have little patience with scientists who take
a board of wood, look for the thinnest part,
and drill a great number of holes where drilling is easy.
—Albert Einstein (quoted by Philipp Frank in "Einstein's
Philosophy of Science," *Reviews of Modern Physics*,
Vol. 21. No. 3, July 1949)

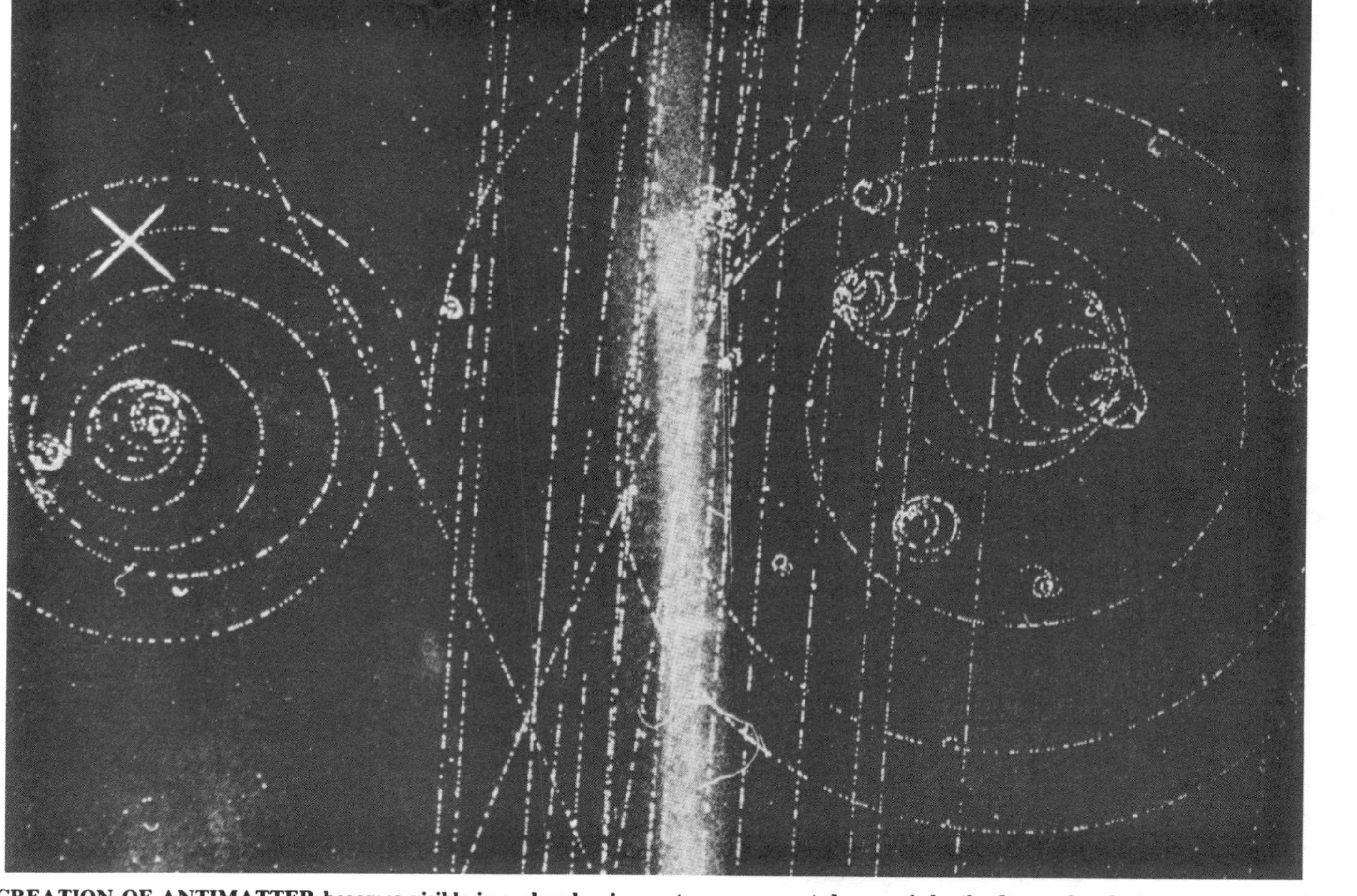

CREATION OF ANTIMATTER becomes visible in a chamber in which the trajectory of any particle that has an electric charge is marked by a trail of bubbles in liquid helium. Here the antimatter is a positron, whose clockwise spiraling path fills the right two-thirds of the photograph. The smaller counterclockwise spiral represents the path of an electron. The positron is the antiparticle of the electron: the two are identical in mass, but in several other ways, such as electric charge, their attributes are opposite. The positron and the electron were created as a pair by the decay of a photon, or quantum of electromagnetc radiation. The photon's path cannot be seen because photons have no electric charge and do not give rise to bubbles in the helium. A magnetic field was applied to the chamber to bend the trajectories of the particles. In high-energy experiments the creation of particle-antiparticle pairs is common, yet the universe in the large appears to consist predominantly of matter. The photograph was made by Nicholas P. Samios of the Brookhaven National Laboratory.

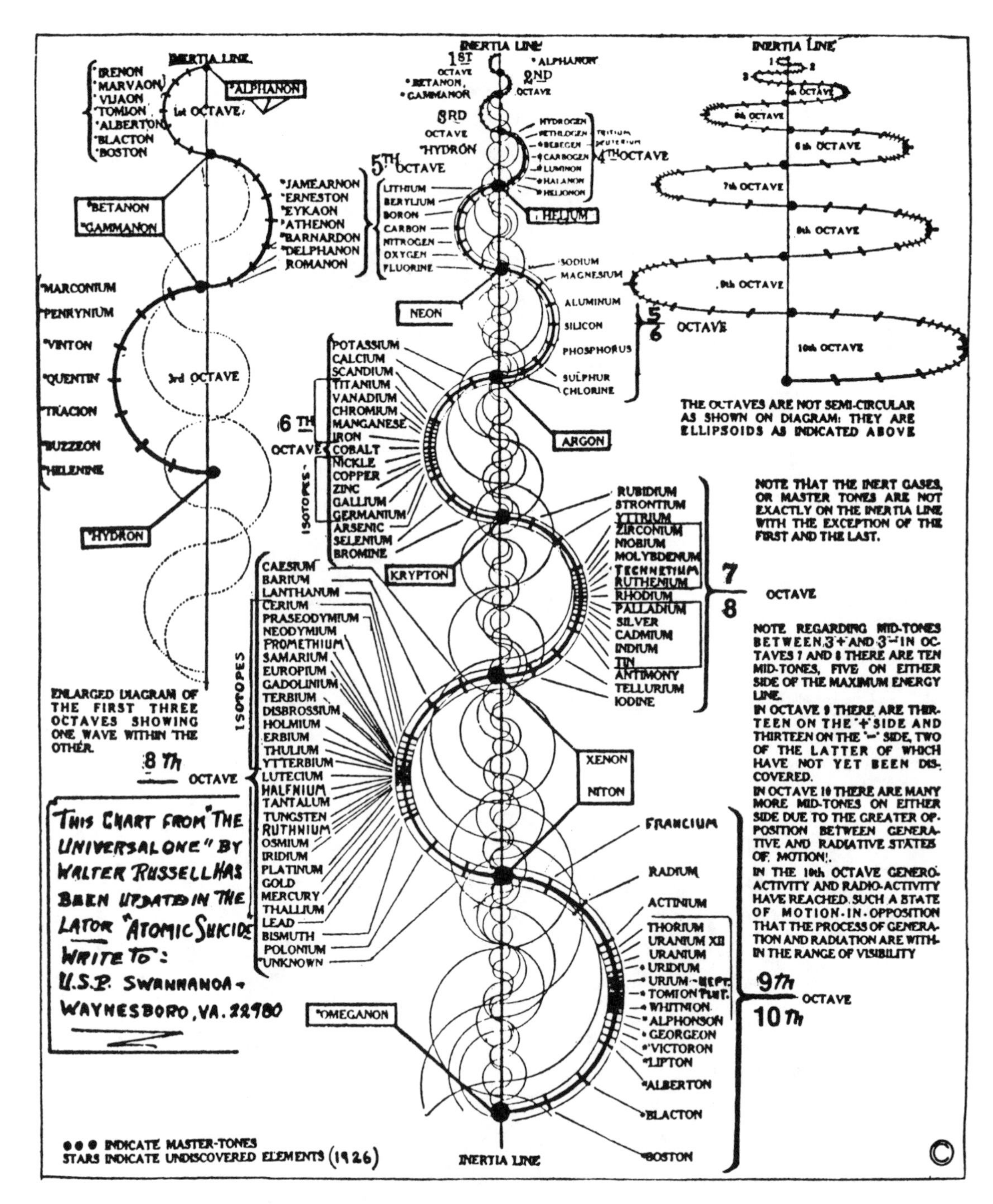

Diagram showing the ten octaves of integrating light, one octave within the other. These ten octaves constitute one complete cycle of the transfer of the universal constant of energy into, and through, all of its dimensions in sequence.

The Unified Field Theory is the most significant physics concept ever conceived, and the implementation of such a theory would revolutionize life and technology on our planet, as it probably has on many other planets.

The relativistic equation $E=mc^2$ stated that the energy contained in any particle of matter is equal to the mass of the matter multiplied by the square of the speed of light, 186,000 miles per second. As this equation implies, even a tiny amount of a matter would release huge amounts of energy, and it is on this basis that nuclear scientists built the first atomic bomb.

In his general theory of relativity, Einstein had treated gravity as a force caused by a gravitational field. Matter gave rise to a gravitational field, which in turn acted on other material bodies to cause forces to act. Einstein had taken this force into account by means of curvature of space. A similar situation existed for electrically charged particles. Forces act between them, and they could be taken into account by considering the electric charges to give rise to an electromagnetic field, which in turn produced forces on other charged particles. Thus matter and gravitational field were exactly analogous to an electric charge and an electromagnetic field.

Consequently, Einstein sought to build a theory of the "Unified Field" which would be a generalization of his gravitational theory and would include all electromagnetic phenomena. He also thought that in this way he might be able to obtain a more satisfactory theory of light quanta (photons) than Bohr's, and derive laws about how our physical reality worked, instead of only laws about observational results.

The great success of the geometrical method in the general theory of relativity naturally suggested to him the idea of developing the new theory within the structure of four-dimensional space. He decided space must still have still have other characteristics besides the curvature which takes care of gravitational effects.

On Einstein's fiftieth birthday (1929) news that he was working on a unified field theory became particularly widespread in the international scientific community. The public at large seemed especially attracted to the idea that on the very day on which he attained fifty years, a man should also find the magic formula by which all the puzzles of nature would finally be solved. Einstein received telegrams from newspapers and publishers in all parts of the world requesting that he acquaint them in a few words with the contents of his new theory. Hundreds of reporters besieged his house, and when a few reporters managed to get a hold of him, an astonished Einstein said, "I really don't need any publicity."

Still, everyone expected some new sensation that would surpass the wonder produced by his previous theories. The press learned that a communication dealing with the new Unified Field Theory would be published in the transactions of the Prussian Academy of Science, and efforts were made by newspapers to secure galley proofs from the printer, but without success. There was nothing to do but to await the publication of the article, and in order not to be too late, an American newspaper arranged to have it sent immediately by phototelegraphy.

The article was only a few pages long and it consisted for the most part of mathematical formulae that were completely unintelligible to the public. The emotion with which the common man greeted the formulae might be compared to that experienced by the sight of an ancient cuneiform inscription. For an understanding of the paper a considerable capacity for abstract geometrical thinking was required. To those who possessed this quality it revealed that general laws for a unified field could be derived from a certain hypothesis regarding the structure of four-dimensional space. It could be shown that these laws included the known laws of the electromagnetic field as well as Einstein's law of gravitation as special cases.

Was Einstein's Unified Field Theory put to practical use and studied by his fellow scientists? According to one of Einstein's biographers, Philipp Frank, "...no result capable of experimental verification could be derived from them. Thus for the public at large the new theory was even more incomprehensible than the previous theories. For the expert it was an accomplishment of great logical and aesthetic perfection."[3]

The Unified Field Theory can actually be traced back to 1860 with James Clerk Maxwell's discovery that electricity and magnetism can be united as the electromagnetic force. Born in Scotland in 1831, Maxwell is acknowledged as one of the finest mathematicians in history, yet it is for his contributions to physics that he is best known. Maxwell was the first to formulate the kinetic theory of gases, and shortly afterward made one of the most important discoveries in all science, leading the way for Einstein and the Unified Field Theory. Maxwell extended the ideas of the physicist Michael Faraday and interpreted them in rigorous mathematical terms. Like Faraday, Maxwell visualized the *space surrounding charged bodies* as filled with lines of force carried in a *field of electricity* and associated a field of magnetism with the space. In a few equations based on this concept of fields, Maxwell devised a mathematical expression of all the varied phenomena of electricity and magnetism. It proved beyond doubt the indissoluble link between electric and magnetic fields, and added the term "electromagnetism" to the vocabulary of science.

Using his equations, Maxwell showed that the oscillation of an electric charge would produce an electromagnetic field radiating from its source at a constant velocity. He worked out the velocity to be 186,300 miles per hour, roughly the velocity of light. He immediately suggested that light was therefore a kind of electromagnetic radiation and that visible light was only a small part of a much wider electromagnetic spectrum. Proof of the genius of Maxwell's work can be seen in the wide uses of the spectrum he described; from long wavelengths of radar and radio waves to ultrashort wavelengths of X-Rays, the laws governing the behavior of radiation are expressed with brilliant simplicity by his equations on the electromagnetic field.

In Maxwell's Treatise on *Electricity and Magnetism*, a scientific work as important to 19th Century scientists as Einstein's *Theory of Relativity* was ours, Maxwell showed that the quantities oscillating in a light wave are the electric and magnetic fields. Maxwell theorized, correctly, that the electric field, as a moving electric charge, was inextricably linked to the magnetic field, as changing magnetic intensity. Maxwell theorized that electric and magnetic fields have peaks of maximum and minimum strength, analogous to the crests and troughs of a wave, but there is no vertical motion. It was Maxwell's equation that overthrew the theory that a light wave oscillated by the displacement of particles in the ether.

Among the ensuing mathematical implications was the odd requirement that whenever an electric field is put into motion or retarded, it must not only set up a magnetic field at right angles to it but must send out (perpendicular to both) an impalpable electromagnetic wave that travels through space with the speed of

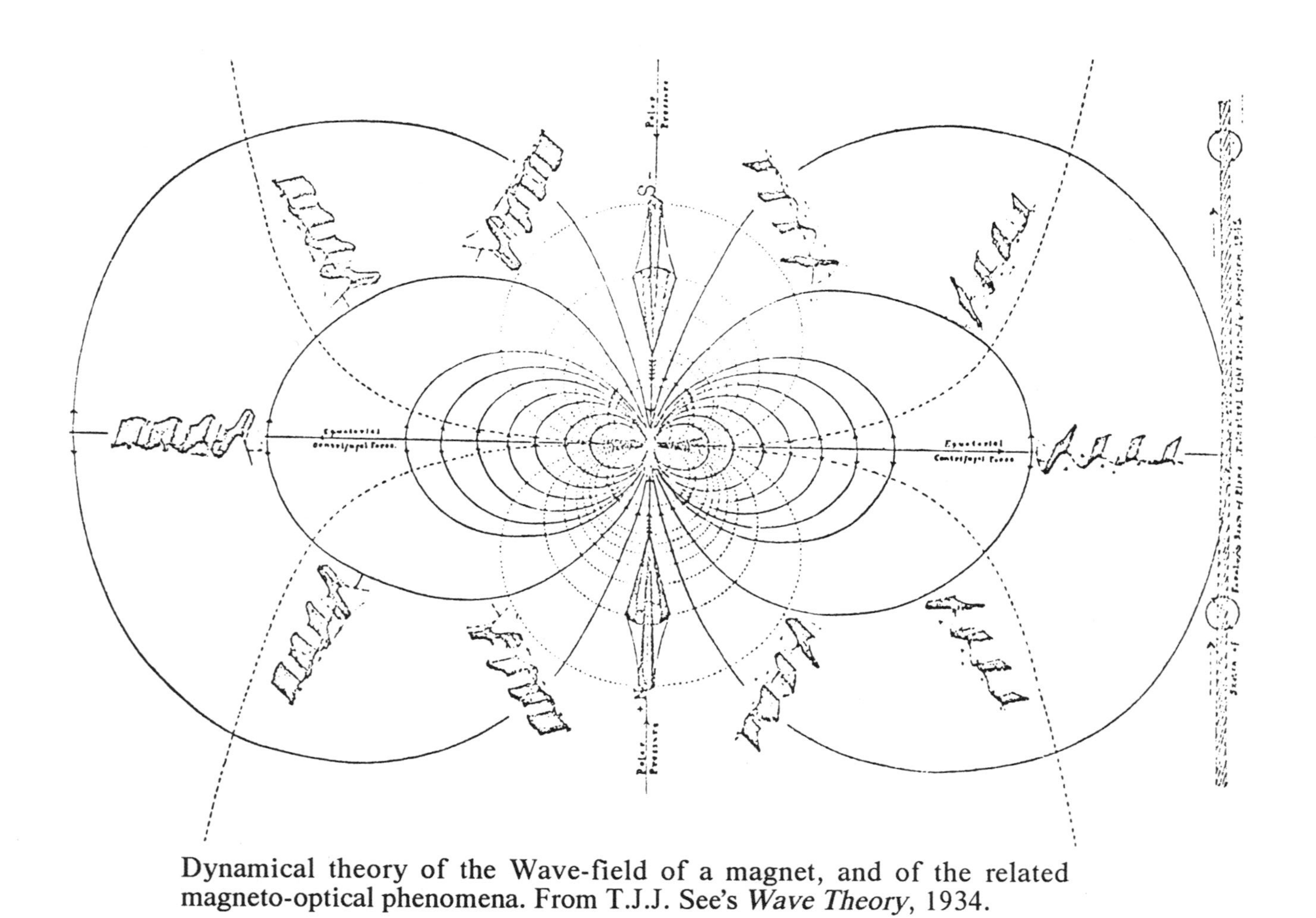

Dynamical theory of the Wave-field of a magnet, and of the related magneto-optical phenomena. From T.J.J. See's *Wave Theory*, 1934.

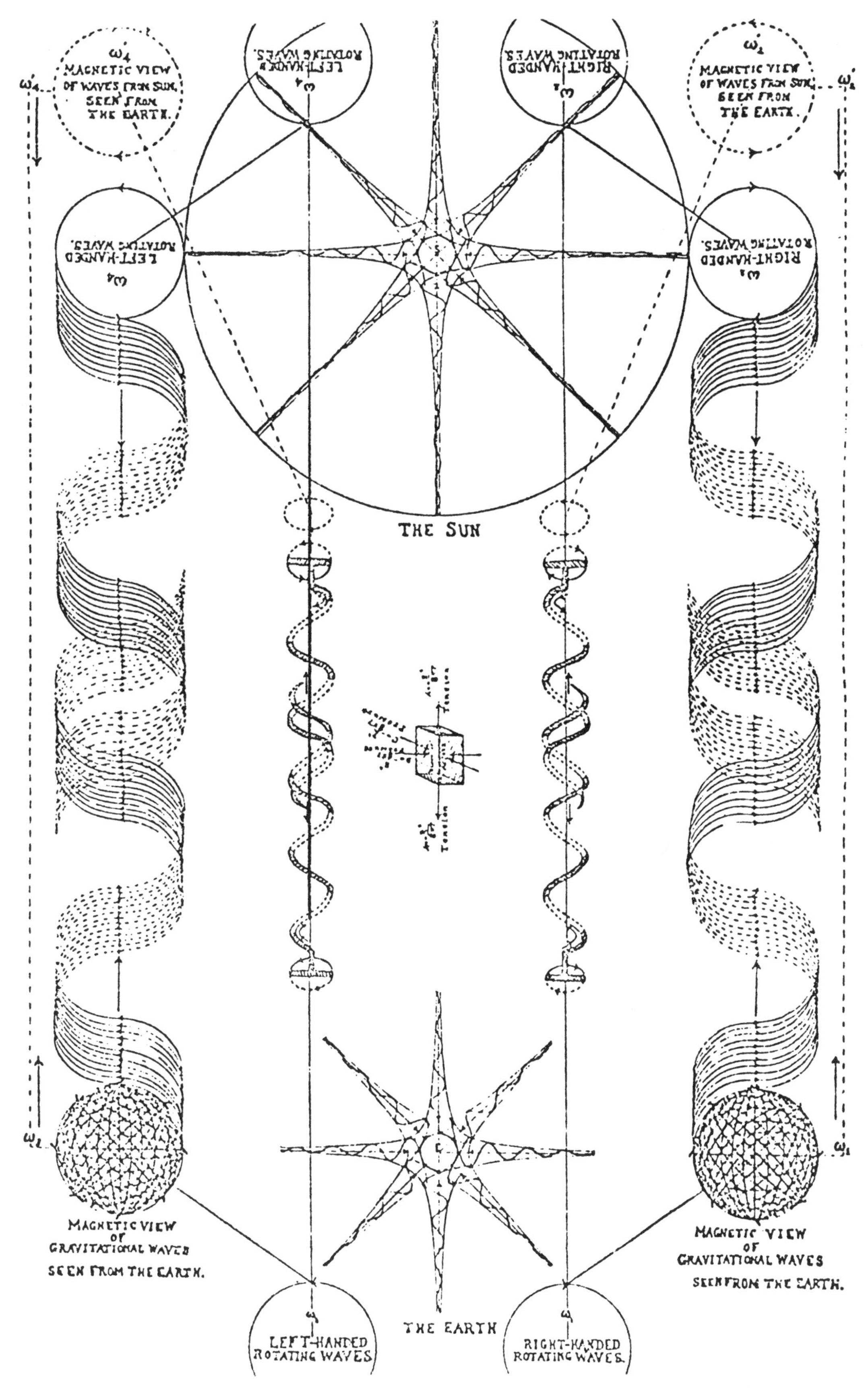

T.J.J. See's diagram of gravitational waves, incorporating magnetism, electricity, gravity, wave theory and the Unified Field. From *Wave Theory*, 1934.

light and has other important properties derived from Huygen's wave theory of light. This unprecedented development struck Maxwell as exciting in its potentialities. Could such electromagnetic waves—if they turned out to exist—really be a kind of light? Was light, in fact, a form of vibrating electromagnetic energy which might have other forms at other frequencies? Maxwell concluded exactly that.

Maxwell was a brilliant scientist, but he was one giant step away from a Unified Field theory. He needed only to add one thing to his list of components of the quantities working within magnetic and electrical fields, as well as light waves. That component is of course gravity.

Gravity, Electricity and Magnetism

"Einstein is not difficult, only unbelievable," said one of his critics.

The link between gravity, electricity and magnetism is usually defined as the basis of the unified field. Whereas Maxwell's theory on electrical and magnetic fields said that the two are intricately interconnected, manifesting from one source, so does gravity also manifest from the same source with similar links to the common source. Similarly, while Maxwell showed that there is an equation for electro-magnetism, there is similarly an equation for electro-gravity, magneto-gravity, and even electro-magneto-gravity. Equations for gravity control, in which gravity is seen as a manifestation of the unified field, would be a portion of a unified field equation.

Electro-gravity and magneto-gravity would be artificial (?) forms of gravity with which gravity could be controlled and warped into fields around aircraft, spacecraft, and other vehicles. A unified field equation that provided a practical development of gravity control would be the single most important discovery since electrical power generation.

Another contemporary of Einstein who worked on the Unified Field was the scientist T. J. J. See who in the 1930s published seven manuals (a total of about 4000 pages) on his extensive research into the understanding of the gravitational riddle. Unfortunately, See's work was largely ignored and is virtually unknown in the scientific community. It has been said that he had the unfortunate tendency to "come on too strongly" with his theories which was usually interpreted by his astronomer colleagues as being "arrogant egotism" rather than "studied conviction."

Actually, See's technical work, and in particular his *wave theory of gravitation,* was quite competently done and is now viewed by many as providing a sound basis toward an eventually accepted Unified Field Theory. The basis for See's wave theory is that Pi is an infinite oscillating series leading to an expanded theory of curvilinear motion. The oscillating series correspond to dynamic impulses — physical waves in the ether, as postulated by Huyghens and Newton for the curvilinear motions of the stars observed in the immensity of space. In See's diagrams gravitational, magnetic and electrostatic fields are presented as longitudinal/compression waves in the ether, of widely divergent wave lengths. These various wave forms are of a proportionate magnitude to the distances over which they operate. Yet despite See's careful research he was ignored then, as he is ignored now, and fifty years later, it seems we are no closer to an accepted unified field equation than we were at his time.

It is interesting that such an equation seems beyond our present day scientists. Even though Einstein first unveiled his Unified Field Equation in 1929, little or no work, has apparently occurred on the subject for the last fifty years! Can this be the truth, or has work on the Unified Field and its promise of artificial gravity continued secretly since the theory was first proposed? Investigators into the Philadelphia Experiment during WWII would certainly answer that question affirmatively. The tense war-time environment of the late 30's and 40's was a time when secrecy was of the utmost importance, and when the profound and startling implications of a new phase of technology made a hiding of the results from the masses an absolute necessity. Yet today, after the war, need we still conduct such research in secret?

With the detonation of the first atomic bombs, and the end of the war, a new form of energy was released in a devastating way on the Japanese people as well as on an amazed and unsuspecting American public. Suddenly, the very fabric of the Universe seemed to have been tapped, and the energy released was awesome in its nature. The unified field theorists now had another force to contend with and work into part of the equation: The Nuclear Force.

In the book, *Superforce*[10] by the British scientist Paul Davies, the author says that the interactions in the universe can be shown to be the consequences of four basic forces: gravity, the electromagnetic force, and the two nuclear forces, strong and weak. All four are vital to our present existence.

Gravity holds us to the earth and the earth to the sun. The electromagnetic force holds atoms and molecules together; manifested as light, it enables us to see. The strong force holds the atomic nucleus together, enabling complex atoms to exist. The weak force controls the reactions on the atomic level that allow the sun to shine.

In the last decade or so scientists have found that the fundamental particles these forces act upon are quarks (with names like up, down, strange, charm, truth and beauty) and leptons (electrons, neutrinos, muons, tauons). Of the old fundamental triad we learn in high school—neutron, proton, and electron—only the electron survives as a fundamental. A fundamental particle appears to have the amazing property of being a true mathematical point, indivisible, with no spatial dimensions; yet it can have properties like mass, electric charge, spin and qualities whimsically referred to as color and flavor. Having no size, these particles can be squeezed together to arbitrarily high densities, as they are in black holes or hypothetically were in an early universe.

With the frontiers of high-energy particle physics and cosmology, the superforce is used to explain what the "big bang" that gave birth to the universe, and how the foundations for the cosmic structures we now see were forged in the microseconds that followed. It also suggests that space and time are in reality eleven-dimensional, with the unseen dimensions of space masquerading as nuclear and electromagnetic forces.[10]

Davies discusses what he calls the "first breakthrough" in the unified field theory when calculations showed that the weak and electromagnetic forces may become identical at high energies. The experiments of Nobel Prize winners Carlo Rubbia of Italy and Simon van der Meer of the Netherlands at their lab in Switzerland corroborated these calculations. Using a machine designed by Mr. van der Meer, Mr. Rubbia's theory that the weak force and electromagnetic force may become identical at high energies was proven. They have now attempted to

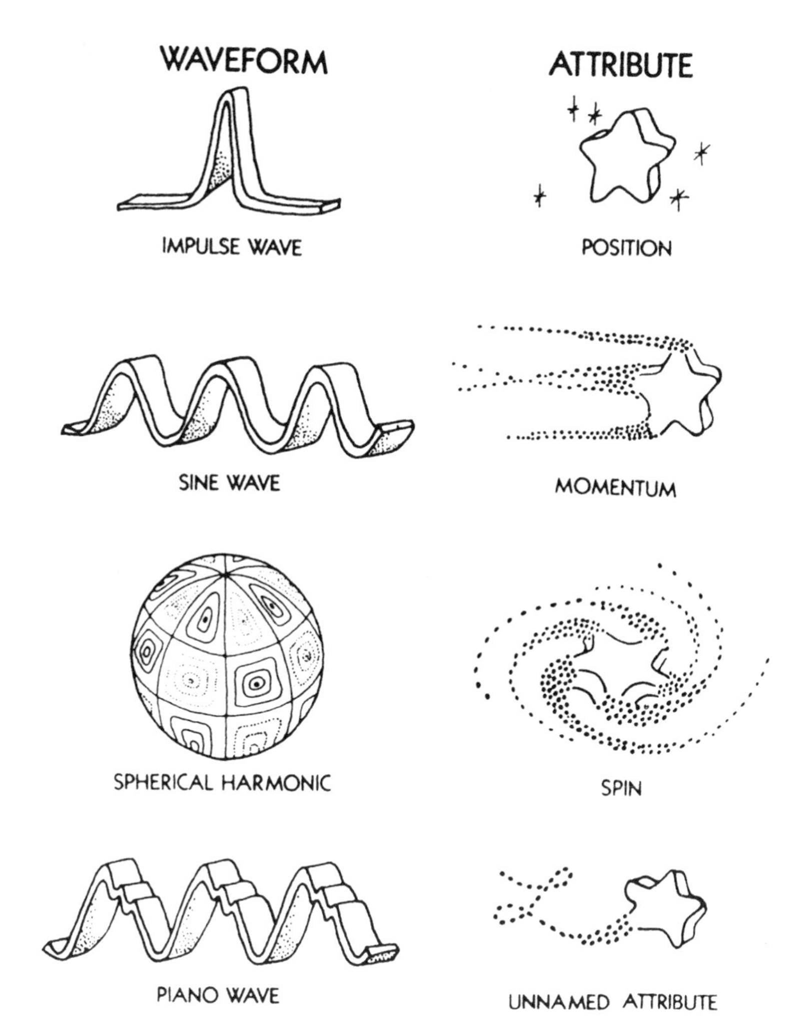

Quantum Waveforms and their attributions. Every waveform family corresponds in quantum theory to a physical attribute — a universal quantum code which is the key to much of this theory's peculiar behavior. Along with each waveform attribute entry goes a rule which connects a wave quantity with a personal name and the magnitude of its corresponding physical attribute. From *Beyond Einstein* [9]

SPHERICAL HARMONIC WAVEFORMS

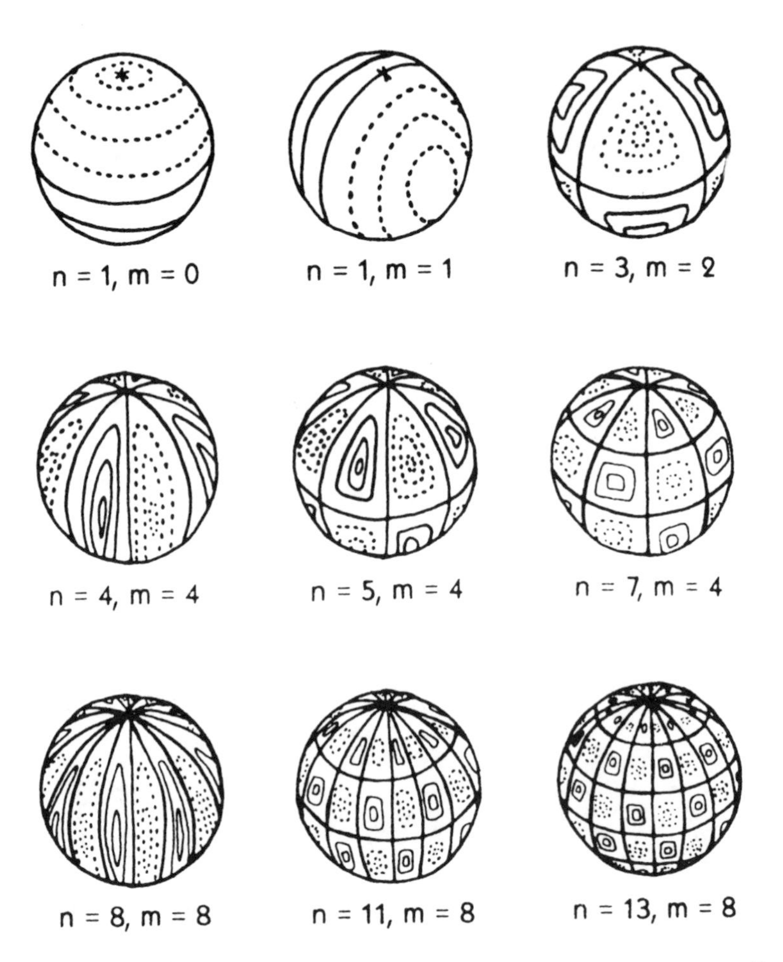

Spherical harmonic wave forms are distinguished by a number ***n***, which counts its nodal circles, and a number ***m***, counting those circles that go through the shere's poles. Sherical harmonic waveforms seem to be the best model for a unified field manifested as a wave form so far discovered. From *Beyond Einstein*[9]

HYDROGEN PROXY WAVES

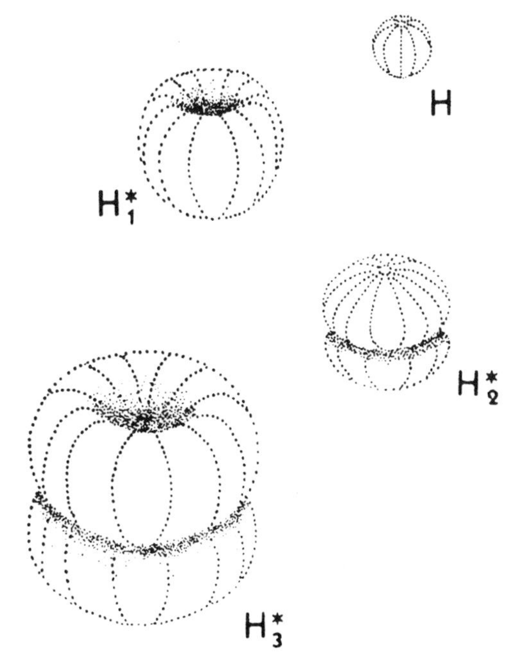

Proxy waves for some states of the hydrogen atom: the ground state (H); the first excited state (H^*_1), the second excited state (H^*_2); the third excited state (H^*_3). Probably the most complete description of a single hydrogen atom, they also show how toriod and vortex rings are the fundamental basis for the fabric of the universe. From *Beyond Einstein* [9]

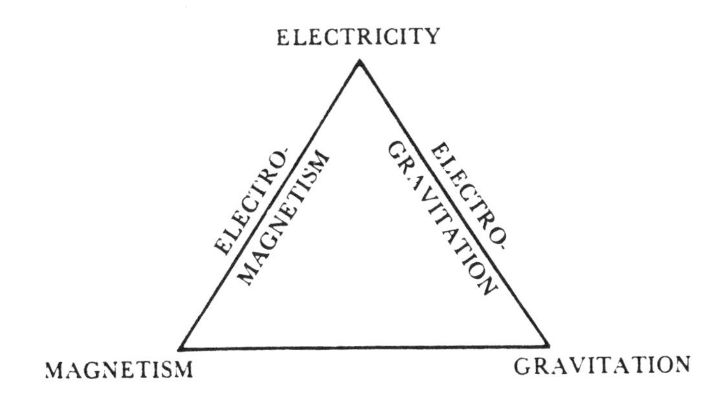

combine the other forces with now unified "electroweak force". Theories that attempt to combine the strong, weak and electromagnetic forces go under the name of Grand Unified Theories (GUTs); a theory combing gravity with the other three is a Super-GUT, probably involving a special symmetry of nature known as supersymmetry (SUSY).[10]

Essentially, all energy and particles must belong to the Unified Field. It is from this field that all matter, mass, wave form, energy and particles are manifesting. Ether is now Quantum particles, and science has proven that there is no void. Energy is everywhere, even in so-called empty space. The outmoded term of ether, supposedly disproved by the famous Michelson-Morley Speed of Light Experiment, now becomes the quantum field, and physics come full circle again, though at least in an upward spiral.

Meanwhile, other energies, such as Wilhelm Reich's Orgone energy, must also vie for their place in the Unified Field. In the Bioenergetic-Orgonomic theories of life and weather, a basic energy, present throughout the universe, flows throughout all living beings and may be part of the solution to orgasmic energy (such as kundalini) as well as the mysterious phenomenon of Spontaneous Human Combustion.

Einstein Discovers the Most Powerful Force in the Universe

The final step in creating a cohesive unified field theory is the addition of one of the most powerful and mysterious forces in the universe. In all seriousness, I conclude that the final, missing part of any Unified Field Theory is Cosmic Love. Though this may sound silly to the empirical scientist, to the mystic this makes a great deal of sense.

Davies concluded that the universe may have as many as eleven dimensions. In most metaphysical teachings, the Universe consists of different planes of reality. In Western Judeo-Christic-Egyptian thought, there are seven planes of existence (God created the world in seven "planes" or "days," as Genesis, the first book of the Bible tells us). According to this concept, the first plane is the physical plane, consisting of what is generally known as physical matter, be it light, a rock or a living organism. The second plane is known as the etheric plane, consisting of electro-magnetic vibration of a higher frequency than that of the physical plane. Auras and fields photographed by Kirilian photography may be of this plane. The third plane is the supposed Astral Plane, in which men and animals are said to have “astral bodies.” The fourth plane is said to be the “Mental Plane” or the plane of “Mind Power.” It is this fourth plane of existence and powers inherent within it that supposedly mark the difference between mankind, with mind power and animals, who lack mind power.

The fifth plane is said to be the next vibrational rate which is the Angelic Plane. Angels are then beings who are of this fifth vibrational rate. The sixth plane would then be the Archangelic Plane, Archangels being theoretically of the sixth vibrational rate. This brings us then to the Seventh Plane of Existence, the “Seventh Heaven” of Biblical lore and the “God Head” or Celestial Host. From this seventh plane it is said that *Love* as a power, an actual energy, manifests.

In this theory, Love is an energy manifesting from the Seventh Plane of Existence, and it is a force like none other. If this belief is correct, we can see how God is literally “Love" and that Love (in the sense of divine Love, rather than the selfish, conditional type of Love most humans believe to incorrectly be Love) is an energy streaming continually throughout the universe as real force. In this belief, Love can then be “channeled” by an individual by simply tuning into the vibration of Love. Love is all around us. Love, a force from the seventh plane, the Celestial Host, and needs merely be tapped.

There seems little doubt that mystics and scientists alike can believe in Love. That Love might be the missing ingredient in a unified field equation is a notion that is bound to tickle a few funny bones, raise a few eyebrows, and perhaps make a few people think.

Perhaps one of the secrets to levitation is that the practitioner must increase his or her spiritual vibration to match that of the seventh plane, pure love, a sort of selfless—surrender to devotion—***energy***. It is this fantastic energy which controls the universe, an energy manifesting from the unknowable Celestial Plane, God himself, the Tao, or that which one cannot known or speak. An energy which has a source, yet is everywhere, and can be tapped into by biological organisms such as humans.

Yet, while we may be able to draw up equations showing the relationship between electricity, magnetism, gravity and the nuclear force, it may just be that Love, as a power manifesting from the Seventh Plane of Existence, will elude any attempts to integrate it into scientific formulae. Yet, while Love remains perhaps the most mysterious of all forces (including gravity), the potential of putting the unified field to practical use in terms of *gravity* control is still within easy reach, if not already accomplished (apparently in secret) by the scientific community. To quote Albert Einstein, "Gravity cannot be held responsible for people falling in love." Indeed, maybe gravity cannot be held responsible, but the unified field can!

1. **Relativity**; The Special & the General Theory, Albert Einstein, 1920, Methuen, London.
2. **Ideas and Opinions**, Albert Einstein, 1954, Bonanza Books, NYC.
3. **Einstein, His Life and Times**, Philipp Frank, 1947, Knopf, NYC.
4. **The Universe and Dr. Einstein**, Lincoln Barnett, 1948, Sloan & Co. NYC.
5. **Einstein, The Life and Times**, Ronald W. Clark, 1971, World Publishing Co. NYC.
6. **Einstein's Universe**, Nigel Calder, 1979, Viking Press, NYC.
7. **Time Warps**, John Gribbin, 1979, Dell Publishing, NYC.
8. **Quantum Reality**, Nick Herbert, 1985, Doubleday, Garden City, NY.
9. **Beyond Einstein**, Dr. Michio Kaku & Jennifer Trainer, 1987, Bantam Books, NYC.
10. **Superforce**, Paul Davies, 1984, Simon & Schuster, NYC.
11. **The Particle Connection**, Christine Sutton, 1984, Simon & Schuster, NYC.

THE VORTEX ARENA
by
John Walker

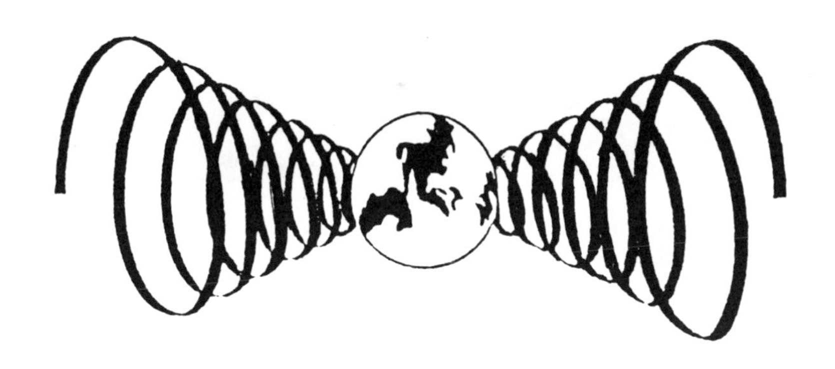

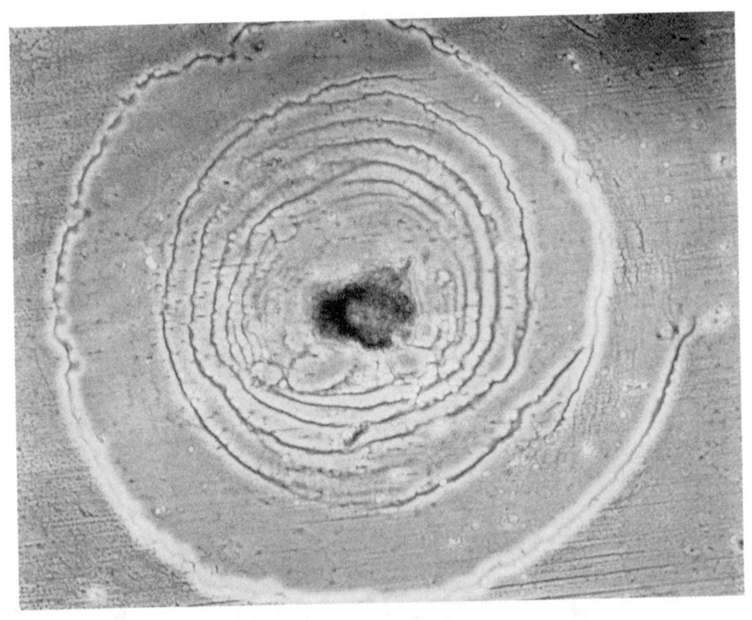

This photo comes from a Sydney University laboratory, (New South Wales, Australia) and was taken in late 1986. Regarded as a historic event, the photo shows the low-energy spiral field that surrounds a carbon particle. The particle was placed on a polymer-coated slide. Polymers are used by scientists as a substitute for human cells. They can act as sensitive film emulsions and thus record an impression of a physical event — in this case, that a previously undetected field pattern surrounds this carbon particle. Energy is believed to flow in spirals. This photo is considered to be evidence in support of the *Unified Field Theory*, which Einstein, amongst others, believed to the basis of the physical universe. Ancient Hindu tradition says that this life force, known as Sakti or the 'energy of the gods', materialises in the form of a spiral and brings with it the forces of attraction and repulsion. Photo and caption courtesy of Mark Balfour, from his book ***Sign of the Serpent*** (1990).

THE VORTEX ARENA

Vortex Power In Ancient India

There is a curious document, translated in the 1950's from the ancient India Sanskrit, to English, the "VYMAANIKA—SHAASTRA" or "SCIENCE of AERONAUTICS" a document written in the 4th Century BC, though taken from other documents thousands of years old, describing *aircraft* of that time. We have here a kind of pilots manual for the operation of at least four types of craft called Vimanas. In the contents are outlined air routes, airplane parts, types of metals for the crafts, modes of power and interestingly enough "WHIRL-POOLS" or vortices the pilots should be aware of. This entire document appears in the Adventures Unlimited Press book *Vimana Aircraft of Ancient India & Atlantis,* including comments by Sanskrit scholars.

To quote:

> "Aavartaas or aerial whirlpools are innumerable in the above regions. Of themthe whirlpools in the routes of the Vimanas are five. In the Rekhapathha there occurs "Shaktyaavarta" or whirlpool of energy. In the Mandala-pathha there occurs the whirlpool of winds. In Kakshyaa-pathha there occurs Kiranaavarta orwhirlpool from solar rays. In Shakti-pathha there occurs shytyaavarta or whirlpool of cold currents. And in Kendra-pathha there occurs gharshanaavartaor whirlpool by collision. Such whirlpools are destructive of Vimanas, and have tobe guarded against.
>
> The pilot should know these five sources of danger, and learn to steer clear of them to safety."

In ancient India the writers of knowledge were careful to observe every form of change, every pattern of flow-rest-motion and to describe even the smallest of effects seen, the causes unseen. Often they spoke of matters that were beyond the five senses, yet in much detail. It seems their science was one of experience more than speculation.

Indian Knowledge

The Indians of North America fashion Medicine Wheels, Kivas and sacred circles, sometimes with a stone or stones clustered in the center. The walls of the Hopi Kiva are round. During the sacred snake dance, the participants recite words which reveal that the Kiva is far more than a stone circle or wall on the surface, but that it is the *connection point* with inter-related forces both above and below the Earth line. The dance circle of the American continent goes in one direction while on the other side of the globe the native dances are of the other direction. Dance direction is determined primarily by sensing what is already present, a vortex or vortices. Unseen whirlpools.

Sometimes the Dance is to *create a focal point* where there was none before, and sometimes the Dance is to break up an undesired pattern, for balance between the forces of nature in the sky and Earth. The Kiva according to both the Hopi and the Zuni Indians, extends into the Earth as an *inverted energy cone.*

Hopi elders believe that there is a seeing of things that can't be explained. There are shrines out there in the spiritual center which are markers for spiritual routes that extend in all four directions to the edge of the continent. They believe that through their ceremonies it is possible to keep the natural forces together. They believe that their prayers go to all parts of the Earth. These are the sacred places, to disturb them is to create an imbalance in the spoke of the "Great Wheel".

As I understand, many of the Hopi shrines, the focus of their prayers and rituals have been desecrated. The delicate network of power paths which radiate out over the Mesa land helping to keep energies of the planet harmoniously stabilized have been interrupted and damaged. I ask the reader, how can something we can't see affect something as large as this planet ? How could the soldiers marching around the walls of Jericho for days, stop, blow their trumpets and bring a fortification down to rubble ? A type of UFO is seen to pass low over the water and witnesses say the water peaked up toward the bottom of the craft. The water was *swirling* and it peaked UP! What is the cause? How could the craft create an upside down waterspout?

We will take a look at much more than this and it will become clear what all of these things have in common, in fact the Unified Field is what we are standing on inside the Vortex Arena.

Definition and Assumption

By the opinion of Mr.Webster a vortex is, "a whirling motion of a fluid forming a cavity in the center—a whirlpool, a whirlwind." Well, if we picture the *ether* as a kind of fluid, and we hold that gravity flows from the outside to the center of a particle, planet, or whatever—then a tornado is kind of like the waters of ether going down into the drain of the Earth. Gravity is just a name for a flow. Don't let the name pull you down, it's taking us with it (or trying to) on its way to the center, pushing through us as it were.

One time a lot of people thought this was what was taking place but somebody somewhere said it couldn't be that simple. I myself thought Newton was pretty clear on it when he stated to the effect, "and the apple falls AS IF it were attracted to the Earth from within." He didn't say "BECAUSE," he said "as if" and today scientists act as if he said because. And boy does that make for some math headaches. *That mass pulls with its own gravity , is an assumption .*

By the way, I might point out here that the "ether" IS what matter IS NOT. Sometimes ether is called "space" or "void," "virtual-state," even "zero-point energy," but that cannot change the amazing fact that one cubic centimeter of fluid ether, contains enough vibrating raw *potential* to be the substance of a galaxy—

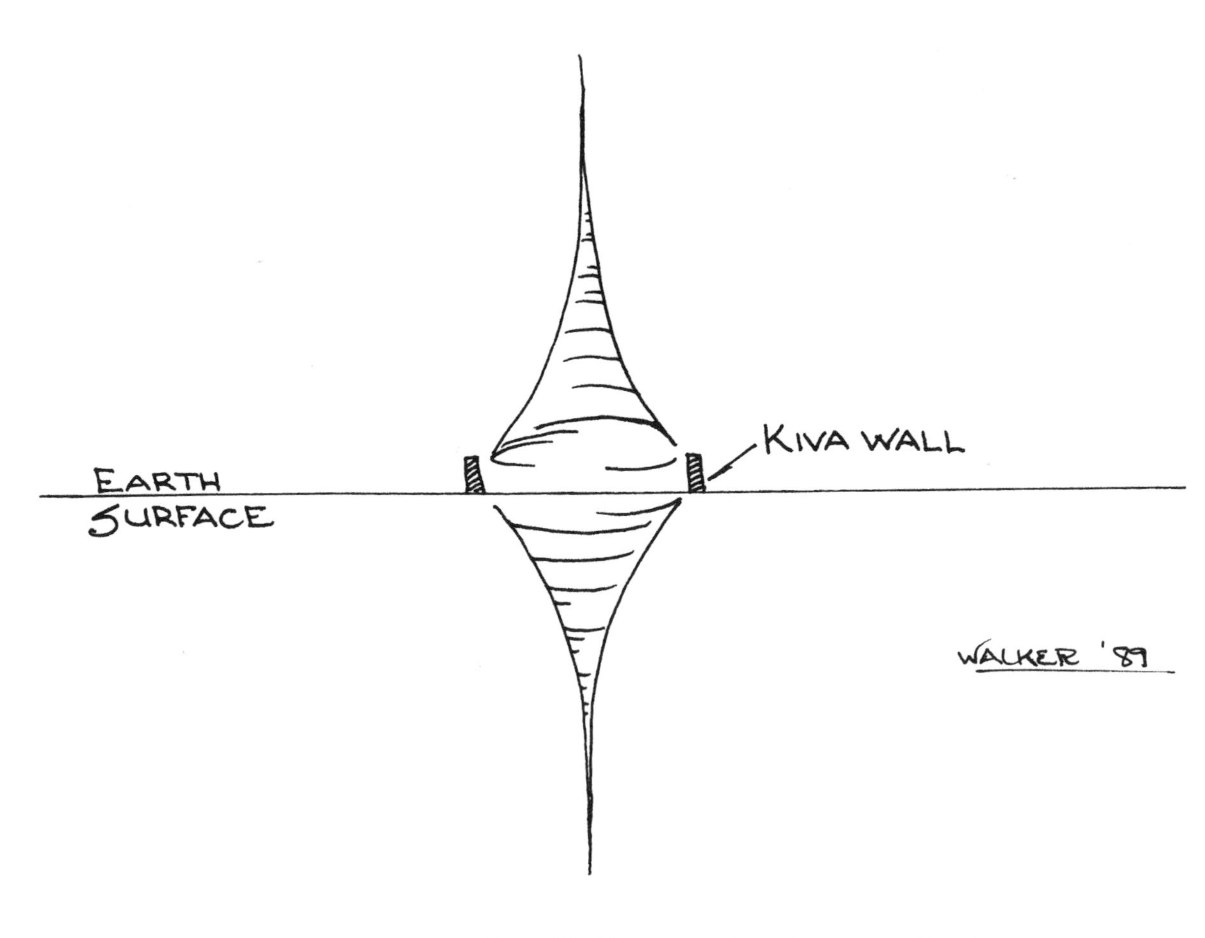

By placing round walls or KIVAS at certain points on the Earth the Indians create sacred places to send their prayers through.

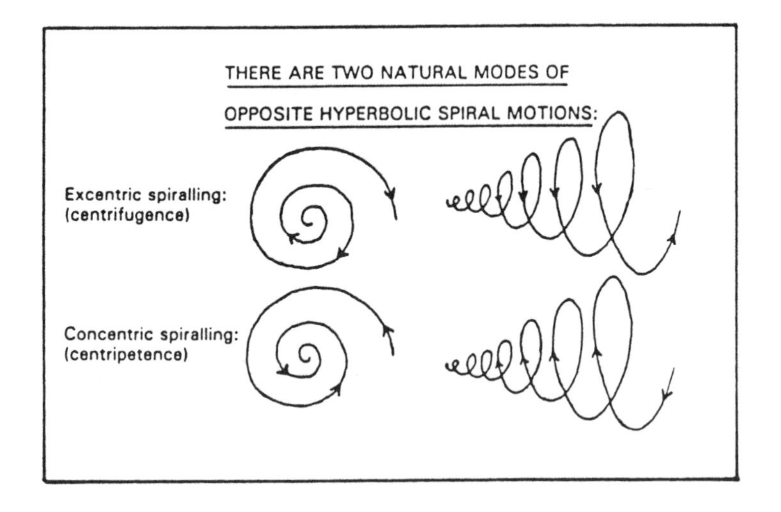

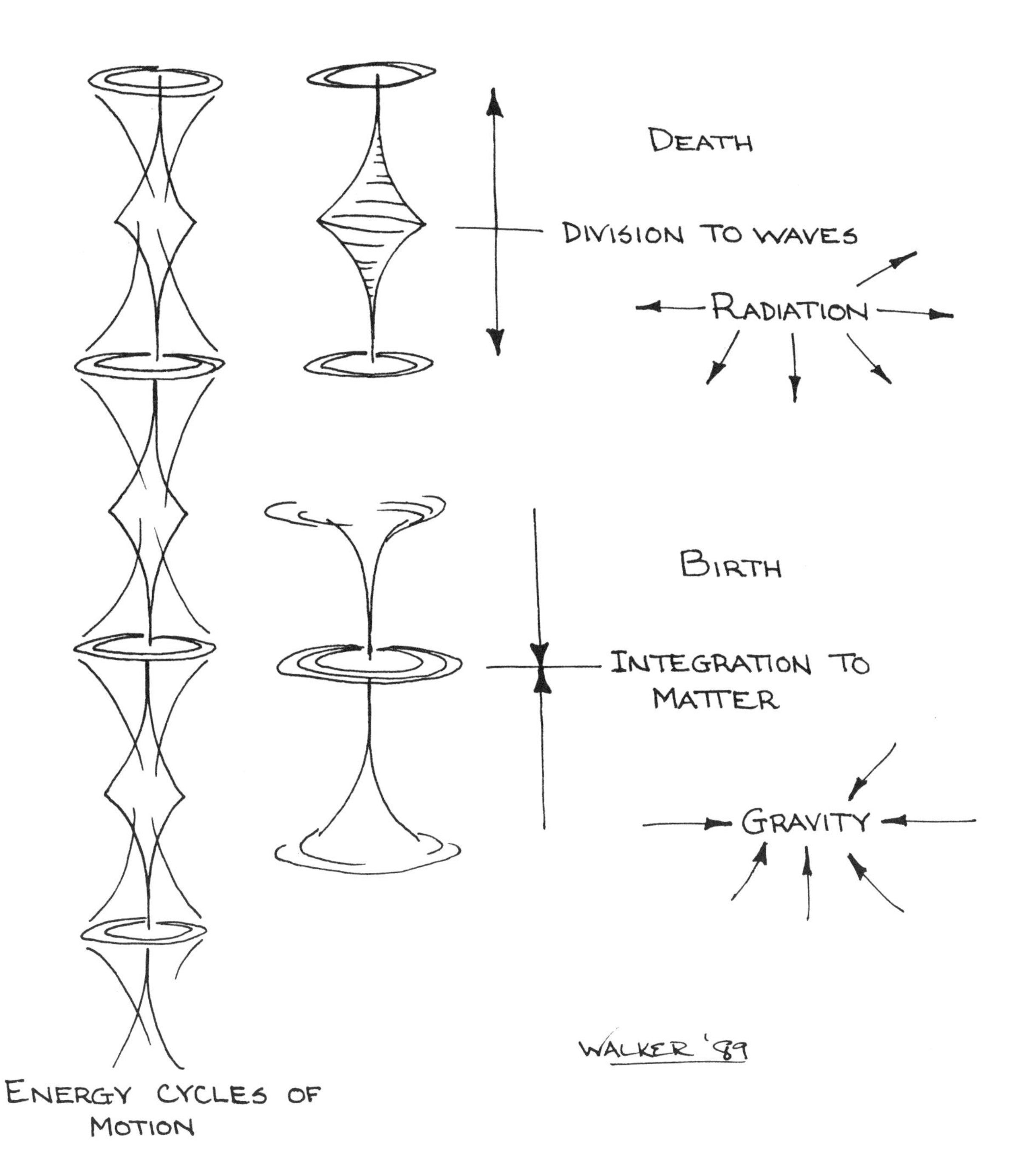

Energy and mass are interchangeable and can be treated as such in mathematics.

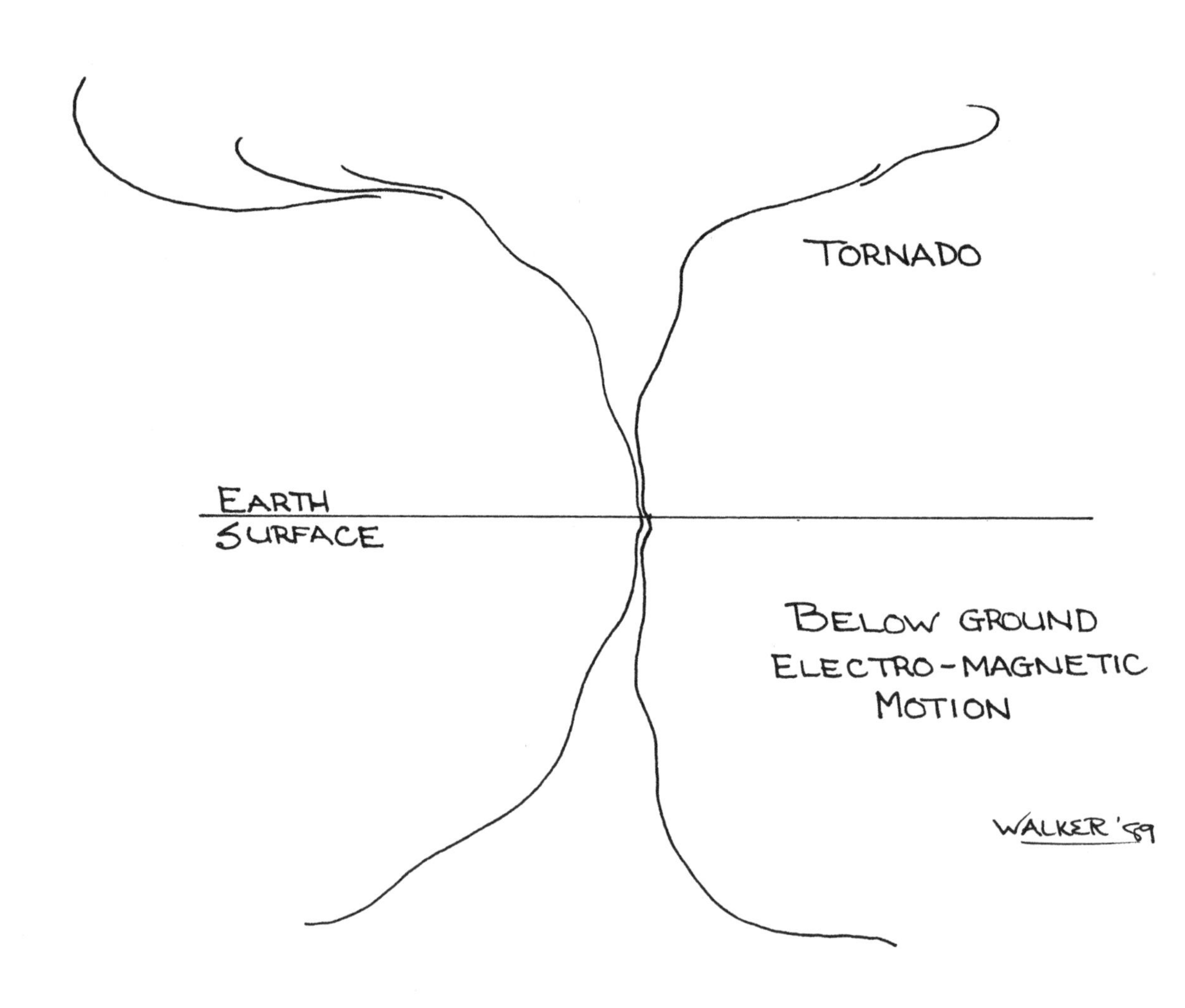

A tornado has its effects below the surface as well as above.

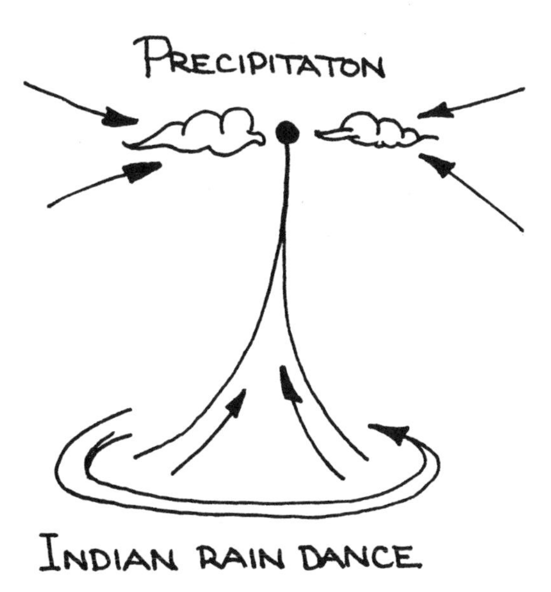

WALKER '89

Rain Dance - By dancing a specific way in a circle, the Indians produce a focal point which precipitates clouds above them. Rain is the result of their efforts.

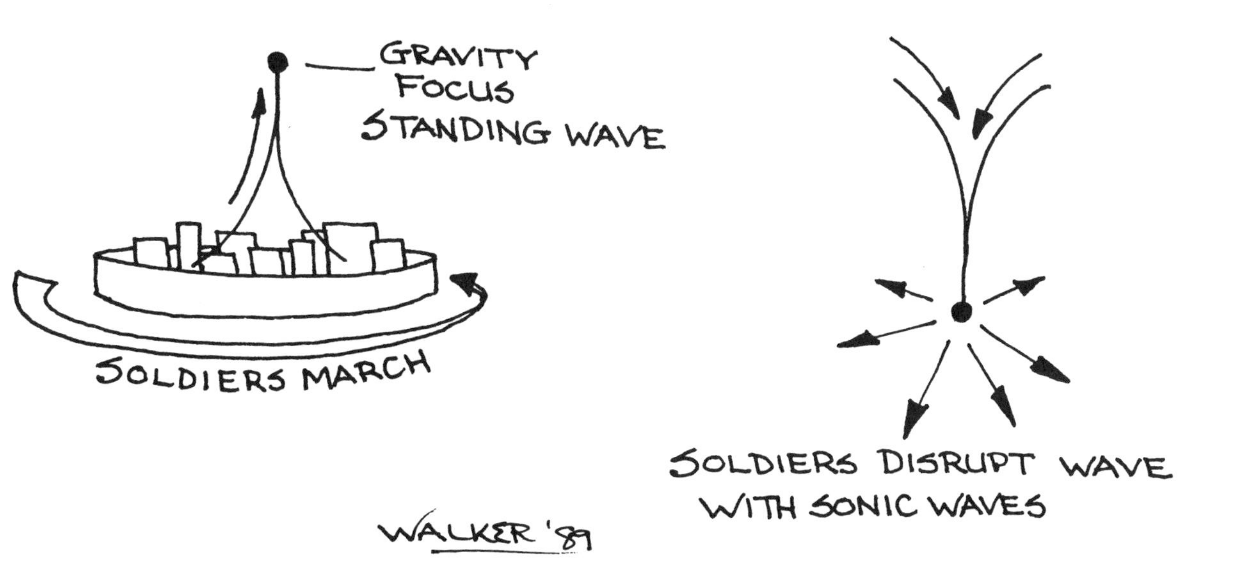

This may well have been how the soldiers at Jericho brought down the walls.

planets and moons thrown in - no extra charge.

You may say, "but there's nothing out there...", and you're quite right for our senses cannot perceive complex waveforms or vibrations so quick in their coming and going that it APPEARS still and at rest. I will refer to this "stillness" as the *gravity axis* around which all motion as we know it, circulates.

The primary substance *ether* has a natural tendency to flow in a spiral vortex manner *to the center* of whatever IT supposes to be matter in-the-making, and it spirals OUT from the center as *radiation* in its own "death" process. The localized birth - death - birth cycle of a galaxy the size of an atom, expresses its "time line" in just fractions of a second. The relative birth - death - birth cycle of an atom the size of a galaxy expresses its "time line" in billions of man years.

Now I'll briefly re-introduce you to Nikola Tesla who had some interesting words about this ether/vortex stuff, maybe he would help me here.

Tesla Discusses Ether

In a New York Times article dated April 21, 1908, pg.5, col.6, entitled "How the Electrician's Lamp of Aladdin May Construct New Worlds", Mr. Tesla is speaking of man's mastery of the physical universe by adopting certain theories and (Italics mine) he states:

"Every ponderable atom is differentiated from a *tenuous fluid*, filling all space merely by spinning motion, as a whirl of water in a calm lake. By being *set in movement* this fluid, the ether, becomes gross matter. Its movement arrested, the primary substance reverts to it's normal state."

The normal state here is one of "stillness" where radiation then goes after it's time line as matter. He observes:

"It appears, then, possible for man through harnessed energy of the medium and suitable agencies for *starting and stopping ether whirls* to cause matter to form and disappear. At his command, almost without effort on his part, old worlds would vanish and new ones would spring into being. He could alter the size of this planet, control its seasons, adjust its distance from the sun, guide it on its eternal journey along any path he might choose, through the depths of the universe. He could make planets collide and produce his suns and stars, his heat and light, he could originate life in all its infinite forms. To cause at will the birth and death of matter would be man's grandest deed, which would make him the mastery of physical creation, make him fulfill his ultimate destiny."

Thank you Mr.Tesla!

Gravity Derivative

This is a good time to bring to your attention another factor to consider about this vortex arena we're walking around in, and that is the Earth's local vortex expressions we've labeled electric/magnetic, causing the researcher to see these as two separate entities when in fact it is closer to give this a "first derivative" (first observable) effect from a singular *ether-as-gravity* in motion. Thus the electric/magnetic, or in physics terms, the "H" vector (electric receptance in negative OHMs), and the "E" vector (EMF of magnetic induction, lines per second) are the subdivided offspring from a common parent, AND these "H and E" Siamese twins occupy the SAME space at the SAME time within geometrically wonderful 90 degree angles. By *same space* I must clairify they intersect and occupy locally the same space, and yes it is relative to higher or lower derivatives - hurricanes, galaxies, chakras (which means spinning wheel), a plant sprout coiling up out of the ground, water down the drain . . . look around, the vortex motion is in everything.

Not only that, somebody sometime, built these pyramid and stonehenge "enigmas" fully understanding the motions involved so they could play around with this effect, while we're *outstanding* in our fields, watching the grass grow.

Let's use our second sight and observe what is being created during an intense emotionally charged sacred Dance.

Life is motion, and motion arises as a derivative of gravity-come-ether. Now I state quite emphatically that what ever moves here physically, be it your hand through the air, underground stream of water, aircraft flying by — is creating through motion a kind of spiral coil around it as long as it moves. These ARE NOT air currents but rather *currents of the electric potential* (your body is compressed electricity) and the *velocity of the potential,* superimposed to produce what is known as a vector.

If you play with the idea of producing standing waves in a vector and superimpose it with other standing wave vectors, then you have one kind of the pump-wave / push-pull gravity axis effect that Tesla mentioned.

As Above So Below

The Indians don't need all this math stuff because they knew maybe from the past or simply *intuition* that when you go around in a circle on the Earth you create a standing wave, a kind of stationary vortex rising *up* into the sky and equally going *down* into the Earth. You see, the Earth acts as a sort of mirror at the surface so that what ever is extended up whether a physical tower or electric wave, its other or opposite is extended down as well. When you walk in the woods your *negative body potential* is walking right on your heels, upside down. Some dowsers I've met seem to be aware of this fact and being aware reinforces the whole thing.

Does this imply that what exists or is buried in the ground, that then this is mirrored above the ground as a positive potential even if we can't see it ? I confess I assume this is so. Back to our friends the Indians. If our friends know anything about Ley Lines and Earth Grids which they indicate by the Four (actually there are many more than four if you include the minor directions, but four - North, South, East, West symbolically covers the foundation of overall understanding) Directions, then they also know that to produce something *vertically* on the surface, taps into something under the surface *horizontally* at - you guessed it - 90 degrees. They are locked in. This is the so-called right or correct angle. Are we Unifying yet? We can demonstrate this whole 90 degree aspect with grade school physics to see how the vortex principle applies from the *smallest* to the *greatest* of motions in matter.

We thread a wire through the center of a sheet of paper and hold it taught up and down. While this is held vertically, use your third hand to adjust the paper to roughly half way between both ends, and horizontal or 90 degrees to the vertical.

This paper in our example is our ETHER PLANE or FIELD upon which we randomly sprinkle iron filings which represent unseen complex, quickly moving *wave patterns.* There is no order here as we can perceive, however there IS vast potential, so while your wondering about this your friend hooks up the two ends of the wire and turns on the juice. Now there is *motion* going through the wire (and your body if you insist on pulling it tight with your hands) and as I've said, all motion — in this case through the wire, produces a spiral coil. How do we know?

Look down at the paper from the top of the wire. The filings by simple experiment have become orderly in circular fashion around the wire. If you evenly space several sheets or planes along the length you'll find the circle motion goes the entire length, tapering or becoming smaller at the ends where the motion, the electricity, enters or exits the boundry of the wire. You can let go of the wire now. You'll notice also that the filings are scarce inside the circle close to the wire. Why? Because we've just made a *miniature hurricane*, and as everyone knows, the eye of the storm is calm.

This is an important realization for it is relative to all that would have motion. For motion to exist, there must be a centering factor. This factor is the gravity focal point that *waves as gravity* move toward.

I won't get into the particulars of the wire getting warm because its atoms are trying to explode *outward* from the

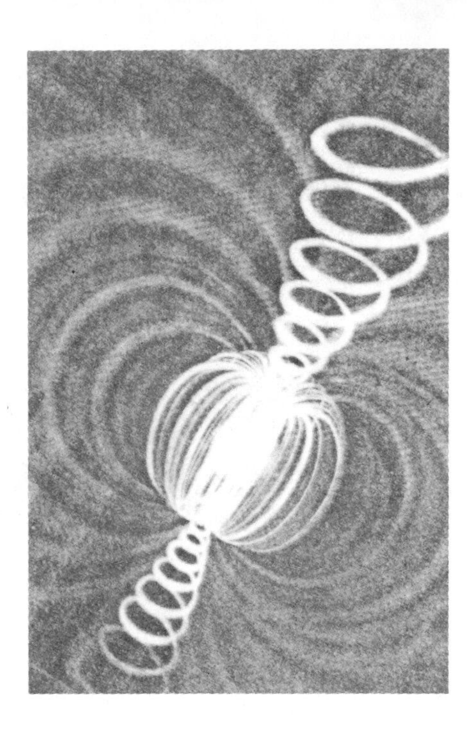

This closely illustrates what we would see, as wave energy became a particle of matter.

BELOW - Two planetary spheres act as focal points for gravity waves. When in proximity to one another, the local gravity is buffered somewhat between them. The result is what <u>appears</u> as attraction.

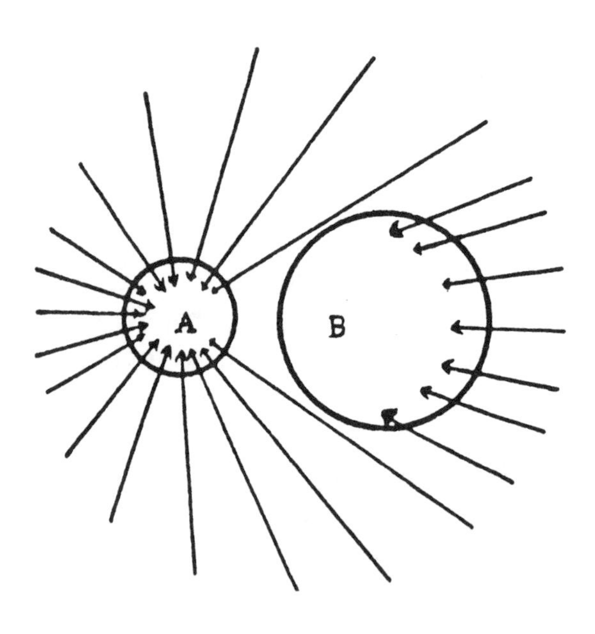

(sketch source: Kinetic Space and its Speculative Consequences by Dr. R. M. Manley)

This photograph of a natural gravity vortex, known as a tornado, can be reproduced on a smaller scale in the lab.

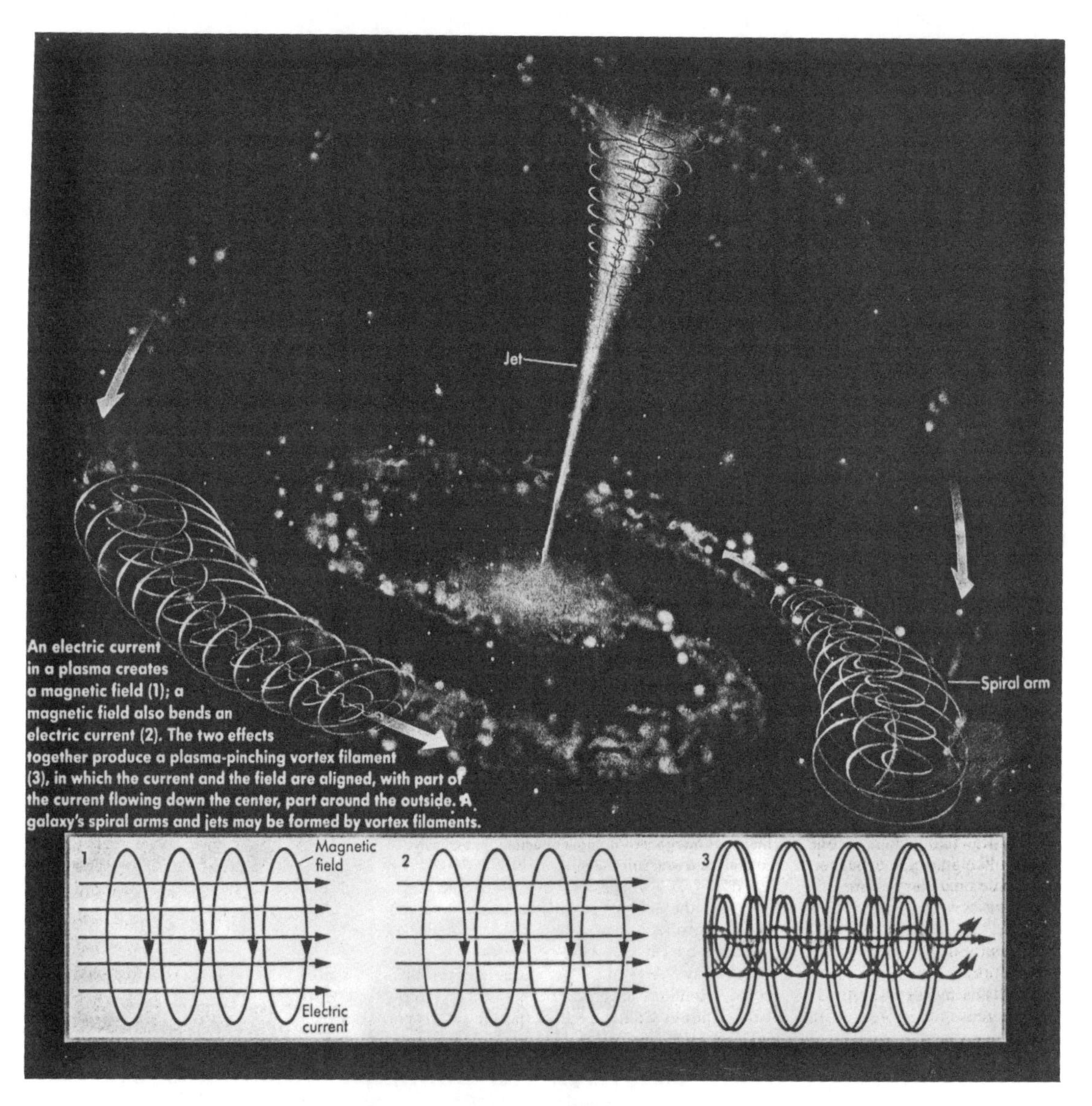

Jet
Spiral arm
An electric current in a plasma creates a magnetic field (1); a magnetic field also bends an electric current (2). The two effects together produce a plasma-pinching vortex filament (3), in which the current and the field are aligned, with part of the current flowing down the center, part around the outside. A galaxy's spiral arms and jets may be formed by vortex filaments.
1
Magnetic field
Electric current
2
3

Scientists are becoming more aware of the part that spirals and vortices play in the creation of matter from atoms to galaxies.

center of its "eye", put enough juice into it and it will radiate its particles out in the form of low level (infra-red) radiation that our senses record as heat and light. Remember the toy tops that spun faster as you pumped the handle up and down? That's what is happening in our wire but the spinning is an electro-magnetic field composed of the electrons of the wire that *in a more restful state* are close in to the wire. The wire "dies" centrifugally and outward where the radiations of what was "solid" wire, now join the ether field as waves. What they were *before* they became particles on this great particle we call the Earth.

In all of nature there is the motion, the calm, and the difference of potentials that *seek to balance* through motion. Isn't to be centered — to be calm? Doesn't the human eye have a *hole* in it surrounded by color? The center is the axis of the cosmic wheel.

Legend or Explanation?

One day by the suggestion of their leader, some soldier guys set out to this place called Jericho. Most of them were probably wondering why they were going to march around this place as they were used to fighting like heck to get anything done. Well this leader got the message to cause a great axis potential to rise up from the center of Jericho and from the circular plane of their marching the potential would rise.

You see, you can pump the axis directly from the top or bottom like a verticle wire, OR you can pump the axis from the outside with a ring of motion.

So they marched and caused a great buildup over Jericho. Invisible, but dense and precipitating as an inverted tornado peaking toward the sky. At this point the inhabitants of Jericho must have felt lighter, for *gravity itself had taken a focus* far above their heads.

The leader also got the message that if they did all of this and then just walked away, the great buildup would gradually subside and they'd be outstanding in their field watching the grass grow. THE TRUMPETS! Yes they brought along their trumpets, no not for Mardi-Gras, but rather to cause a sudden disruption in the flow. Well they blew their hearts out. Two things that have an effect on any and all matter is gravity and sound waves. The leader gave the message to Jericho, the rubber band snapped back to Earth equalizing to its former state of rest, and the soldiers watching all of this from the outside had something to write home about, although I doubt they called it; "Local gravity stress production followed by spontaneous anhilation via sonic reverberation in the direction of the local gravity base plane." Besides, who would've believed them ?

More than Theory

There is a book entitled "Reality Revealed — The Theory of Multidimensional Reality" published in 1978 (Vector Associates) which explains logically matters such as pyramid energy, electricity, gravity, polar reversals, psychic phenomena and many other unexplained phenomena by stating in the forefront that under certain circumstances, the local reality of one truth can be extended and overlap the local reality of another. Of this I have no doubt and I am fairly certain that since that writing, the authors by name - Douglas Vogt and Gary Sultan, have done considerable more in their work and have had new cognitions on the subject.

What they have written regarding tornadoes and hurricanes is to be commended for they are "*getting it* " where others have failed to see. I myself have cognitive experiences where I suddenly "get it" followed by closet cleaning of the mental kind. Thus I advise, leave the doors to your mind open for frequent sweeping.

Doug and Gary refer to a "cross talk" effect where time and space information from one local are transmitted via standing waves of a sort, to another local where the overlap then "unfreezes" into so-called normal time and space.

Standing waves, they observe, can act as a high potential transmitter requiring very **low power** to initiate the action.

In my mind I see a zone of causes and effects transcending time and space when within the local of vortices great and small. The tornado passes and we are left with straw inbedded in unshattered glass, a 2 x 4 piece of pine penetrating 5/8 inch steel, a fifteen inch tire circling the base of a tree whose branches exceed fifteen feet, and metal pipe *under* the Earth is left twisted in the wake of such funnels.

Clearly something other than physical force *as we know it* manifests itself when *conditions* are met. I quote from "Reality Revealed":

> "We contend that the tornado and the hurricane are examples of a cyclotron in reverse." Here the reverse would be a high velocity (motion) creating a high frequency radio (wave) signal. They continue:
>
> "At certain times of the year when the right temperatures exist, a giant capacitor is created. The earth is one plate and the upper atmosphere is the other plate. The earth's magnetic field envelopes these electrostatic plates. We theorize that when the earth is tilted at just the right angle, high-energy-charged particles are actually able *to enter the earth's magnetic field from space.*" (All italics mine.)
>
> "Because of the earth's tilt angle in reference to the direction of the high velocity particle, the particle is *siphoned down* to the tornado belts or hurricane areas. Here the right atmospheric *conditions* exist to form the electrostatic plates.
>
> "The high energy particle charges the plates of the capacitor and a damped, oscillating radio wave is created. High voltage *standing waves* are also created.
>
> "The damped, oscillating wave, along with the earth's magnetic field produces cyclotronic action of the atmosphere. In other words a nature-made cyclotron is created.

Does this crop circle in England trying to represent the Unified Field in symbolism?

Albert Einstein conferring with naval officers in his study at Princeton, New Jersey, July 24, 1943. (*National Archives*)

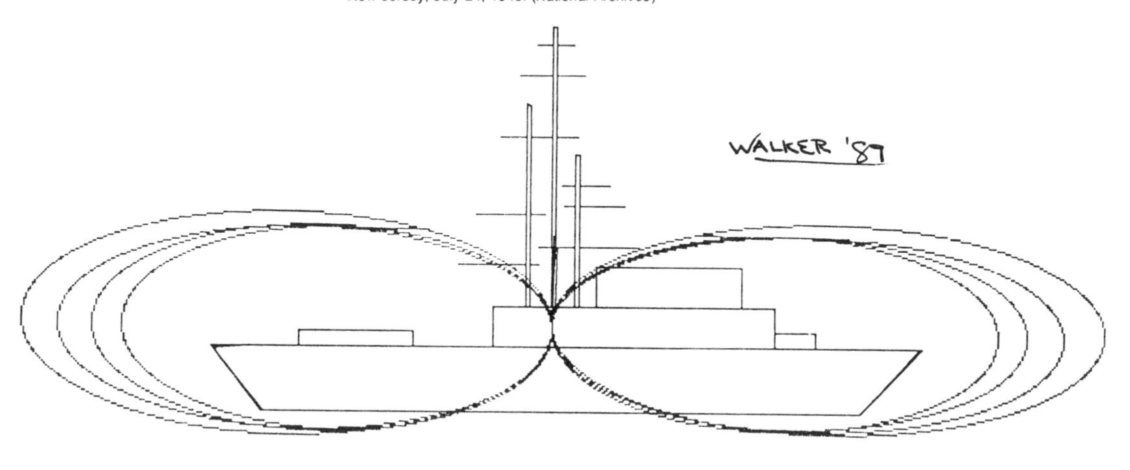

This illustrates the cross section of a torus energy field rotating around the horizontal plane of the ship.

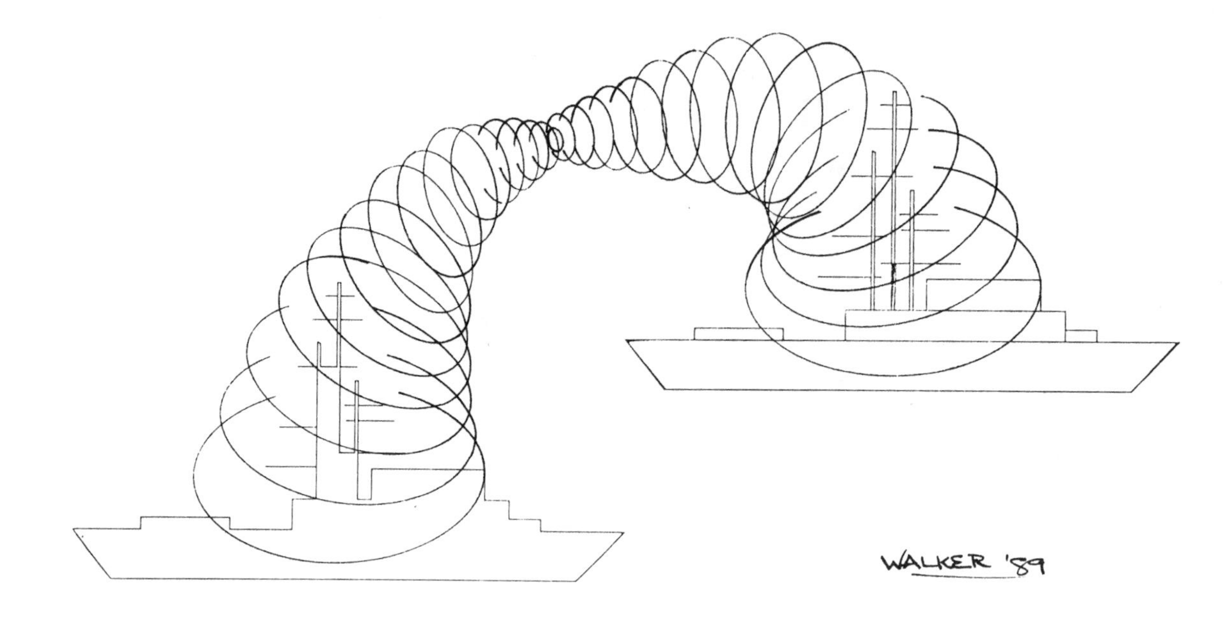

The whole ship and crew would have traversed space as a local electromagnetic tensor field wave dynamic.

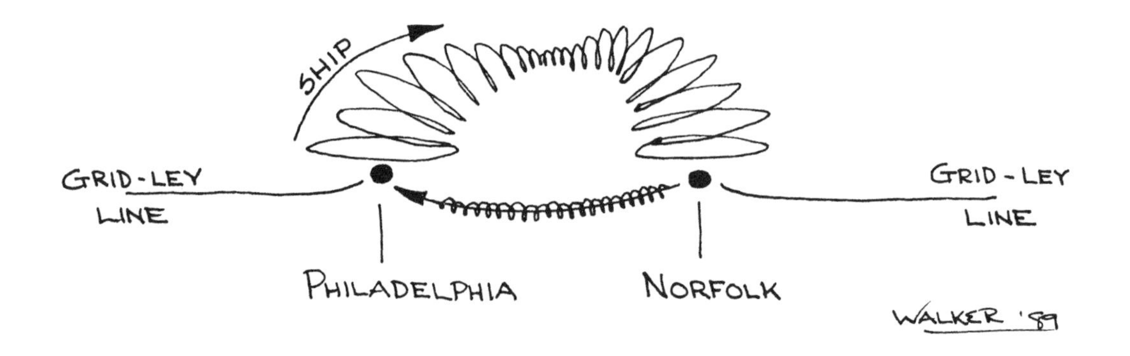

This illustrates how the ship, the DE 173, could have made its 200 mile jump as energy along a grid/ley line.

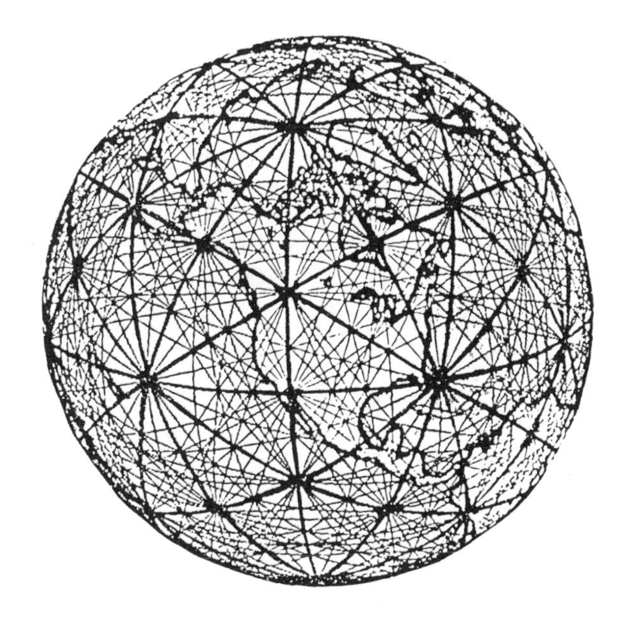

Grid and ley lines are actually similar to spiral ropes of major and minor electromagnetic density. Where they meet or cross there is a standing wave, vortex - like upshoot or downshoot of energy.

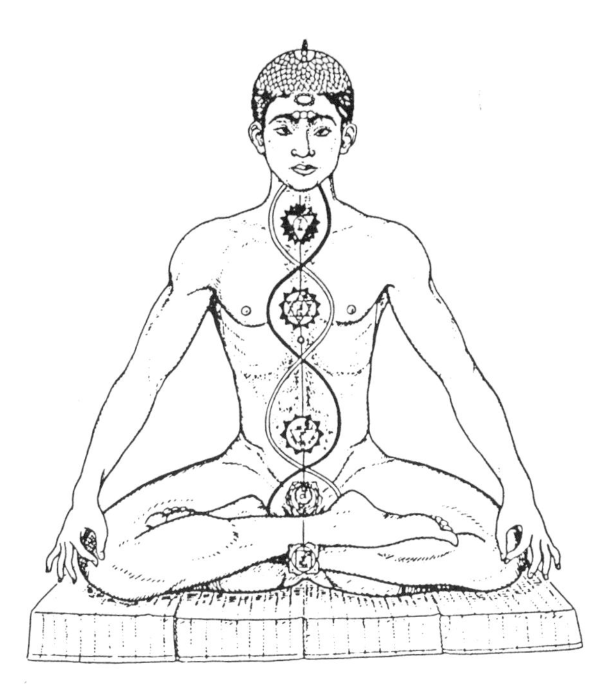

Channels

1. White spiral: *pingala nadi*, solar
2. Dark spiral: *ida nadi*, lunar
3. Central line: *sushumna nadi*, neutral

Spiral motions exist through-out the human body.

"Because of the high electrical potentials created and the high standing waves produced, *information that makes up the straw* is translated (moved) in time; a translation in time is a translation in space. A translation in time of a microsecond of information at the speed of light represents a space translation of about 985 feet. Of course, if the straw just happens to be moved to the space occupied by a window or steel I-beam, it appears as if it has been blown through these objects. But what really happened is that the straw and the I-beam occupied the same space, but at a *different time*, when the tornado was present. When the tornado passed by, the time for the straw and I-beam *became one (unfreezes)* again. The same thing can happen to *living things.*"

We will see shortly how living things and "frozen time" became one in a historical, but little publicized event.

The authors speak of the upper atmosphere and the Earth acting as a giant capacitor in much the same way Tesla referred to it. The idea is not new but I contend that wherever there exist potential differences, there is a flow to balance that difference and this flow expresses its swirling nature. That this expression is the unifying factor between wave and particle physics. And that duplication and application of the unity factor leads to gravity control, teleportation, transmutation of elements, and time travel to name a few.

Simple Experiment

In the mid-sixties there was an article in *Scientific American* about a young high-schooler who submitted an unusual science fair project. The faculty were at a loss to explain exactly what was taking place, and I'm not even sure the young man knew for certain.

The young man built a large simple capacitor — two plates, 8 inches in diameter (the plates were round), and they were spaced 6 inches apart one above the other. To one of the plates — I believe the top one — he applied a high potential voltage in the thousands. On the bottom plate, was a free rolling steel marble or ball bearing. Both plates were smooth and held in place. Yet when he turned on the voltage, the ball rolled around the perimeter of the bottom plate *continually* and *without falling outside the edge* as conventional physics say can't happen.

There are at least two things that took place here. The ball followed the circular flow from one plate to another as best it could, and the ball was held in a field or zone not too close, yet not too far from the center. The ball had become part of the overall effect. Kind of reminds you of planets around a sun does it not? Better still, the ball rolled inside a circular (unseen) corridor whose plane was at right angles to the vertical vortice.

Had a second ball been rolled past this local field effect, it would have been caught up or repelled away depending on the setup.

A Navy Experiment

The U.S. Navy experimented with variations of the unifying vector on a large scale during W.W. II, in which they succeeded at least in part to make a ship invisible to radar. Radar is composed of ultra-high frequency waves, and here enters the problem. In order to make something invisible to ultra-high waves, you must take it beyond that, and now you are playing with an idea where time, space, and conventional geometry go right out the window, literally. The only other way is to create interference vectors *between* yourself and the outside, a kind of shell as it were.

The Navy, Einstein and a few others working together tried it the first way. They subjected the whole ship and crew inside a huge man-made field effect, then pumped its plane-rotational velocity, and raised the overall potential to beyond what might be considered the "normal physics stability" threshold. Then came the book, followed by the movie "The Philadelphia Experiment". Yes somebody out there is trying to tell us something, or make it so obsurd that it falls into the entertainment catagory.

It is here that I'm compelled to shed light on the nature of experimental minds crossing paths with unknown territory. Simply put, the bottom line is a paradox. The risks are great, and unbelievably exciting. Both unknowns until it happens.

In the large field effect experiment, generators were set up to produce a *rotating magnetic field* around the vessel while extra masts were set up to accommodate the *verticle gravity axis* of the field. They had the equipment and some math to back up what was supposed to happen, however it's difficult to prepare human minds to get used to the idea of *transposing* matter from one local to another.

In transit the men on board experienced time distortions and a type of "matter - freeze", non-temperature related, where the men could not move. They would get stuck "semi-comatose" like, where they could breathe, see, feel — yet their world was in their words a "*nether world* ". Interesting choice of words. The crew also spoke of "going blank" and this I assume has something to do with the five senses or possibly going invisible, as opposed to the freeze state.

This freezing, I think, is what we should expect when inside a *hyper-field*, where in all would become coherent or moving in the same direction. You would no longer be a "man energy unit" aboard a "vessel energy unit" upon the sea, but rather the electric fields of both man and vessel would occupy the *same space* at *slightly different times* upon an almost perfect ground plane of water.

There is also some evidence to support the fact that in one of the experiments when ship and crew reappeared, some of the crew had merged bodily within the steel of the ship, just like what we find when a tornado has passed. A rotary motional field is the **basis** for particle-to-wave or wave-to-particle experimentation, for it is the transition point for energy to go either way.

Long after the experiments had ended, some of the crew (that came back) had become insane probably for two reasons. It's a heck of a strain on the brain both physically and mentally. It is also interesting to note here that after the experiments some of the men shifted in and out of our world from time to time, and there were accounts by witnesses of some shifting out and never returning. Apparently their bodies had retained some kind of local hyper-field after affect.

Some of the key statements in the book, "The Philadelphia Experiment" will lead one to similar cognitions if thought about long enough. The authors are William L. Moore in consultation with Charles Berlitz. To quote the book (itallics and underscoring mine):

"As early as 1916, Einstein was busy exploring the possibility that gravity is not really a "force" at all, but rather one of the observable properties of "space-time" itself — the force that underlies and governs all of the other forces in what we consider to be "our" universe. Going one step further, he speculated that what we know as substance, or "matter," is in reality only a local phenomenon exhibited by areas of extreme field-energy concentration. In more simple terms, he chose to view *matter* as a *product of energy* rather than the reverse, and in so doing dared to reject the long-standing concept that the two are separate entities that exist side by side."

Coming and Going

Matter as a product of energy. Substance as a product of waves. Precipitate, gather the waves as in a funnel. The funnel causes rotary motion. Rotary motion causes a funnel. The motion creates a tight magnetic band similar to a torus around the apex of the funnel, the band itself is a spinning toroid which helps to keep what is gathered. All of this is unseen to the human eye. If precipitation continues on its course, the magnetic band becomes the horizontal equator. The vortex funnels become the verticle axis of the poles, of gravity. A particle is born and the principle is now a unit. Now we can see it.

The spin has stabilized and slowed somewhat than at its birth. We can pump it up again, faster and faster the particle will begin to radiate. If we keep this up the particle will "de-centralize," will die as waves of heat, light and radio noise. Yes back to waves. This leads me to believe that the particle is simply a precipitation of wave coherence, that the wave is what is left after particle decay. They are separate only in their time-line of existence.

Seeing is Believing

A man by the name of Carlos Allende or Carl Allen was supposedly one witness to some of the Navy experiments, and part of his own account follows:

> "So you want to know about Einstein's great experiment, Eh? Do you know.... I actually shoved my hand, up to the elbow, into this unique force field as that field flowed, surging powerfully in a *counter-clockwise* direction around the little experimental Navy ship, the DE 173. I felt the... push of that force field against the solidness of my arm and hand outstretched into its humming, pushing- propelling flow.
>
> "I watched the air all around the ship... turn slightly, ever so slightly, darker than all the other air... I saw, after a few minutes, a foggy green mist arise like a thin cloud." (This account closely resembles reports from survivors or observers of disappearances within the Bermuda Triangle, where the aberration may represent a natural — or unnatural — phenomenon on a larger scale.)
>
> "[I think] this must have been a mist of atomic particles. I watched as thereafter the DE 173 became rapidly invisible to human eyes. And yet, the precise shape of the keel and under hull of that... ship *remained*

impressed into the water as it and my own ship sped along somewhat side by side and close to inboards."

Mr. Allende then goes on to describe the sound as a humming which quickly built up to a whispering hum, then increased strongly to a sizzling buzz — a rushing torrent. He continues:

> "The field had a *sheet* of pure electricity around it as it flowed. [This] ... flow was strong enough to almost knock me completely off balance and had my entire body been within that field, the flow would of a most absolute certainty [have] knocked me flat... on my own ship's deck. As it was, my entire body was *not* within that force field when it reached maximum strength - density, repeat, *density* and so I was not knocked down but my arm and hand was [sic] only pushed backward with the field's flow."

Grid or Ley Lines

Clearly this event remained impressed in the mind of Carlos Allende for he was never quite the same again. Also, according to Allende the experimental ship had at one point disappeared from its Philadelphia dock and appeared only minutes later in the Norfolk area 200 miles away. Subsequently it vanished *again* only to reappear at its original dock in Philadelphia. This part of his account may seem incredible until we look at minor and major ley or grid lines that encircle the globe in "The Four Directions".

These lines are actually *very tight spirals* or coils of motion that flow in a more or less linear fashion. Some of these I have found in and around Sedona Arizona to be not more than two or three fingers wide, while others have taken a few steps to walk through. Where these "lines" meet or cross there is an upshoot or a downshoot of motion in the form of a vortex, the likes of which the Indians knew about for hundreds of years.

They also knew how the energy could amplify their thoughts and connect them with the Great Spirit, a concept receiving great attention in the metaphysical arena.

If a physical (to our senses) being or object is vibrated and its frequency is caused to reach certain thresholds, then it becomes a unit of energy capable of following most easily these grid lines around the Earth.

This is what happened to the ship in the experiment. As a unit of energy of high potential, the DE 173 picked the path of least resistance, a line, and gravitated to an area of lower potential — the Norfolk port. Here the ship had lost some of its energy and slowed down, it became visible. However, the power was still turned on, it quickly picked up it's former "carrier frequency" and followed or was pulled back to the original position like a rubber band slowly snapping back from where it was stretched. Researcher Bruce Cathie has determined to his own satisfaction that there IS an energy grid line connecting both the Philadelphia and Norfolk port areas. Mr. Cathie has done considerable in the areas of Earth Grid Networks and Harmonic Relationships.

Unseen World

There is a high concentration of vortex activity in and around the Sedona area but vortex activity itself is not an uncommon phenomenon. There are major vortices in hundreds of areas in the world and minor vortices in countless thousands. We walk through and around them every day. We create them as we walk or drive, and others around us go through our wake. Wind is an EFFECT of the cause which is a high pressure or low sink of electrical potential. The air temperature of hot or cold is simply what we feel as our bodies are subjected to higher or lower electrical pressures or densities.

We see and feel our world in a limited way. Man has always labeled only what he can see, feel, or measure with limited ways of taking measurement.

A simple example would be to place a block of ice at one end of a room while placing a hot iron at the other. Use your mental hologram for this one. If we could see what is normally limited by our perceptions, the cold current would be flowing toward the hot, the hot toward the cold. In electrical terms the voltage is going in one direction, the current in the opposite. Each are high potentials of itself enroute to the "lower side", we only think of it as hot or cold.

Were we to build a very sensitive meter, we could actually measure the electric flow between a block of ice and an iron. It would continue until the ice melted and the iron became cool, or in another sense, when they balanced or became equal in potential. There is something very basic in all of this that I am trying to point out and some of you will pick up on it.

The mass of each is not attracting the other, but there is an interaction that has nothing to do with mass or physical density. Every time you reach for something your hand interacts with the object *before* you actually touch it. There are fields of various energy density, similar to an onion skin, around both organic and inorganic matter which interact continuously. Our senses I assume would be overloaded with information were we to be 100% aware of our environment at all times, however we do experience moments where we know beyond our five senses what is transpiring. The trick is, my friends, to call upon this at will.

The Finger Field

A simple demonstration will show you *something* does interact and you can feel it. Sit back, get quiet and breath easy. With your left hand, bring the thumb, forefinger, and middle finger together so that all three pads of fingers are touching one another, but keep the last two fingers clear of the three. With your right hand extend the forefinger as if pointing and make it orbit the tips of the three on the left hand, but don't let it touch. Concentrate only on the sensations

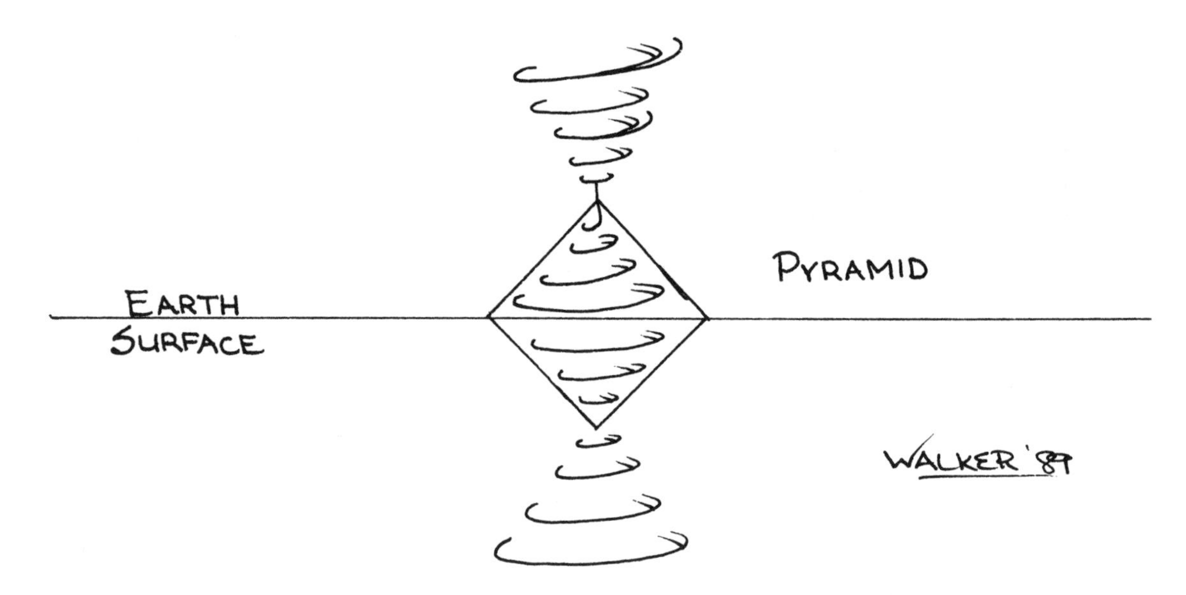

Dowsers can tell you where energy is focussed and which direction it spirals in small scale Pryamids. The large Pyramids are many things to many people, most of which is valid.

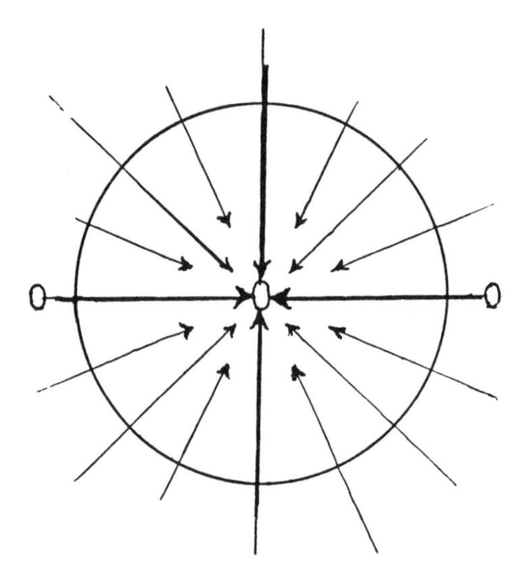

This diagram by Russell depicts gravity waves on their way to the center of a sphere, be it planet or sun.

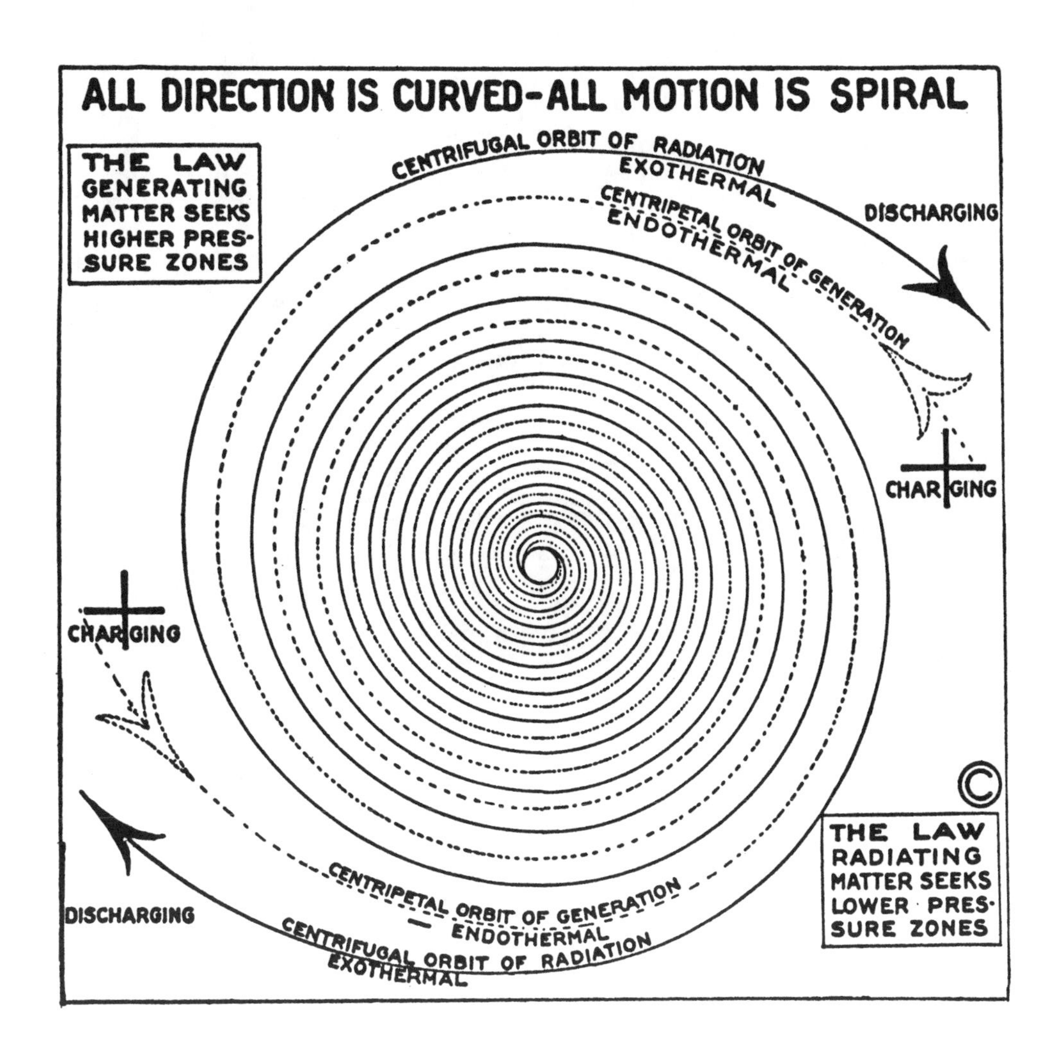

The two basic motions in the universe are the CENTRIFUGAL orbit of radiation - exothermal, and the CENTRIPETAL orbit of generation - endothermal.

By Dr. Walter Russell

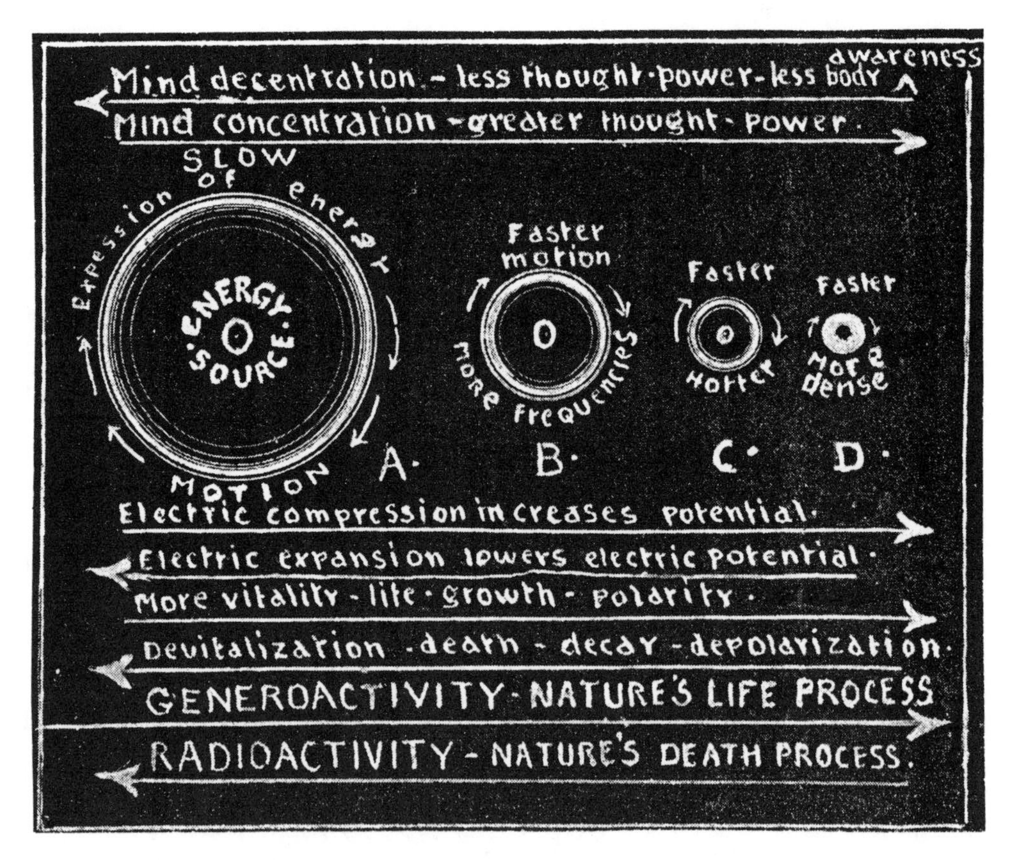

Looking down the gravity axis to the rings of motion.

By Dr. Walter Russell

A. Magnified thought - rings moving around mind source.
B. A sliced section of an electric current, illustrating motion around its Source of energy from which it was extended.
C. An electric current sent through a solid cable of this dimension would be a series of rings, like A and B. It would be a very weak current for there would be a very large hole of non - motion within it.

of the three on the left hand. In a moment you will feel each of the three fingers of the left hand *pulse* as the one right finger passes around. The tips of the fingers may feel to tingle a little. Speed of the rotating finger might be a factor, also you can do this with someone else and the results are similar. You will feel it go away shortly after you stop.

What is taking place? Where is the "field" you're creating, coming from and where does it go after you stop? Welcome to the club.

Open System

In the book, "The Perpetual Motion Mystery - A Continuing Quest" reprinted by Lindsay Publications in 1987, the author Mr. R. A. Ford makes reference to yet another book, italics mine;

"The thought - provoking illustrations described by Gaston Plante in his book "The Storage of Electrical Energy" may be analogous to other natural forces. The *electro - dynamic whirls* experiment was performed with a battery of from 16 to 30 volts. A positive electrode of thick copper wire is immersed in a glass cell containing 1 part sulphuric acid and 10 parts water. Oxidation takes place mainly at the end of the positive electrode. The wire will become pointed in time and the current flow will increase. If the pole of a magnet and the end of the pointed electrode are put close together as shown, a cloud of copper oxide will be seen to move in a quick, *spiral motion* whose direction of rotation is determined by the polarity of the magnetic pole presented.

"These whirls remind us of the Coriolis Effect or a tornado funnel but might they also be connected with such a condition as gravitation? We are accustomed to thinking of a magnet's field as a static condition just as we *assume* gravity to be a condition which has no rotary aspect associated with it.

"We might reason that even though there may be no easy way to visualize how common "inert" materials might be arranged to produce perpetual motion, we must admit that *all materials are ultimately electrical in nature* . At the molecular, atomic and sub-atomic level, frictionless, continuous motion rules the realm."

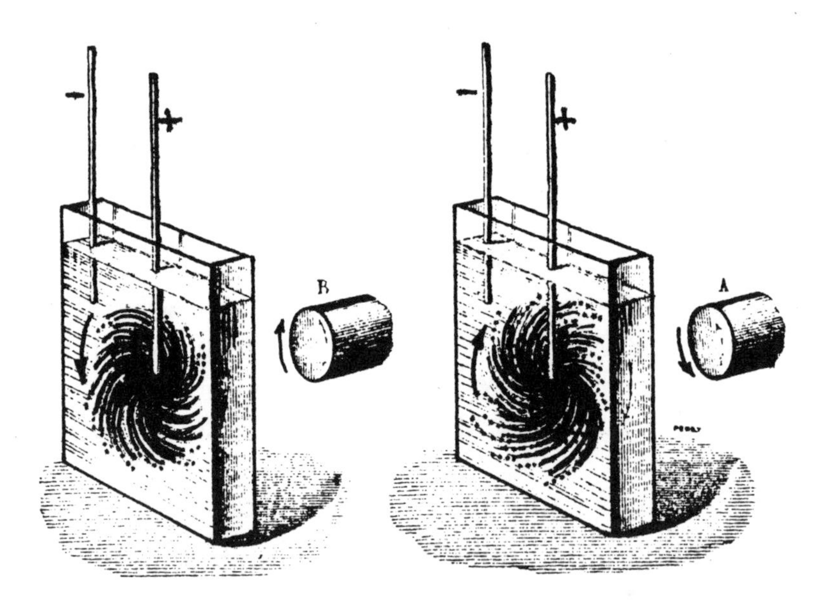

Of particular interest at this point was an American inventer by the name of Charles F. Brush who postulated the wave theory of gravity, known aptly enough as the Brush Wave Theory. In it he states that high - frequency waves of great pressure originate at all points in space, that the energy acquired by a falling object comes from that ether, and is restored to the ether when that body undergoes negative gravitational acceleration. In other words the energy is *borrowed* to produce motion and *given back* when motion ceases.

Mr. Ford's response to the Brush Wave Theory is that; "We are forced to conclude that potential energy, like kinetic energy, *depends* upon motion. It is the *form of motion* which embodies the energy that determines what can happen. From this we see that since the ethereal medium enters into partnership where there is potential energy in a system, there is no such thing as a "closed" or isolated energy system. This implies that there are limitations to the laws of thermodynamics and entropy, which *assume* an ideal, mathematical situation called a closed energy system."

Zero Point Energy

An excellent book for the reader desiring an indepth study of free energy and so called "anti-gravity" is one by Moray B. King, which encompasses 15 years of research tying these concepts to today's modern physics.

The book "Tapping The Zero Point Energy" is actually a collection of Mr. King's papers on the subject, from 1978 to 1989. I had collected most of his published works, but it was a relief to have them all available in one book (this book is available through Adventures Unlimited, see catalogue in the back of this book).

Mr. King states that, "A new technology could await us by combining two heretofore separate, but well documented, areas of modern theoretical physics:

1) The theories of the zero - point energy show there are tremendous fluctuations of electrical field energy imbedded within the fabric of space.

2) The theories of system self - organization might allow this energy to be co-hered by technological means."

I agree with much of what he discusses and told him esentially that an inventor of a free energy device would simply need to give people copies of "Tapping The Zero Point Energy" to explain what the device was about.

Russell Speaks Out

One of my best loved studies on the subject of Science, Physics, and yes even Philosophy can all be found between the covers of a work called "Atomic Suicide?" by Dr. Walter and Lao Russell who founded The University of Science and Philosophy in Waynesboro, Virginia.

This book is hard to find just anywhere, however some

bookstores can order it for you. I've had my copy for over seven years and have read it many times with each session a new experience in understanding is gained.

The human relations of Walter Russell finds him associated with such greats as Mark Twain, Rudyard Kipling, Thomas Edison, and Nikola Tesla, to name a few.

Walter Russell was a master architect, painter, sculptor, musical composer, and as a scientist he was the first to give knowledge of the existence of plutonium, neptunium, deuterium, tritium and many other elements before official science isolated them. Many of Russell's concepts, documentation, and charts were given to the world freely in 1926, however it wasn't until 1941 and three years of lectures did he receive any recognition for his science contributions. He has been called *The Modern Leonardo* by many, and yet the world as a whole is not aware of his great contributions to science and the arts. Such is the walk of genius.

Mr. Russell along with his wife have written much about universal views of nature and human progress. In particular the book "Atomic Suicide?" makes reference to an impending concern for developing new clean sources of energy, and that it is very clear that Nuclear Reactors are not the way to go.

I have studied some about Reactors, enough to convince myself of the long range effects they pose. A Nuclear Reactor has a life of about 25 years, then it needs to be shut down permanently because the very shields that were built to confine the radioactivity, are by that time radioactive themselves. Then the plant area needs to be fenced off to everyone for over two generations. Of the world's operating Reactors, over 50% of them will have to be shut down permanently in five to ten years. This is Nuclear Fission. It heats up water to run steam turbine generators.

What about the new FUSION experiments? Well they still utilize a small amount of radioactivity to produce what they call "cold fusion" and in the opinion of many, radioactivity in <u>any</u> amount goes a long way. Radioactive elements consume oxygen as they decay.

From Russell's "Atomic Suicide?" we look to a brighter future:

"Matter is *motion.* The anti-matter, which now engages the serious attention of science, is *stillness.* Some scientists say it is *pure energy.* If matter, *which is but motion,* is energy, and anti-matter, which is *not motion,* is pure energy, what kind of energy is that which is impure? What does it mean? *The time has come when it is imperative that science must divorce motion and energy as one identity, and regard matter as but the product of the Energy Source.*"

"For scientific purposes in explaining the construction of matter, we will name it (the energy source) the *omnipresent universal vacuum. The universal vacuum is the expansion end of the universal piston, and gravity is the compression end.*

"As electricity is the creator of focal points, which we call gravity, and because compression is the sole office of electricity, every oscillation of the electric current of Creation is an interchange between the stillness of the universal vacuum of God's Mind-universe of CAUSE, and the electric universe of motion, to produce EFFECT."

"It is self-evident that all motion springs from rest and returns to rest for eternal repetition in sequences which we call *electric frequencies.*"

"The greatest thinkers in science have repeatedly said that matter emerges from *space* and is swallowed up by *space* in some unknown and mysterious manner. The word *space* is rather a casual word to use in place of the Creator. Likewise, it is a misleading word, for space is not an expanse of something outside of matter. It is as omnipresent within matter as it is without, and it is in control of matter from within as well as from without."

"Future generations of enlightened men will cease thinking of this electric universe as being matter and

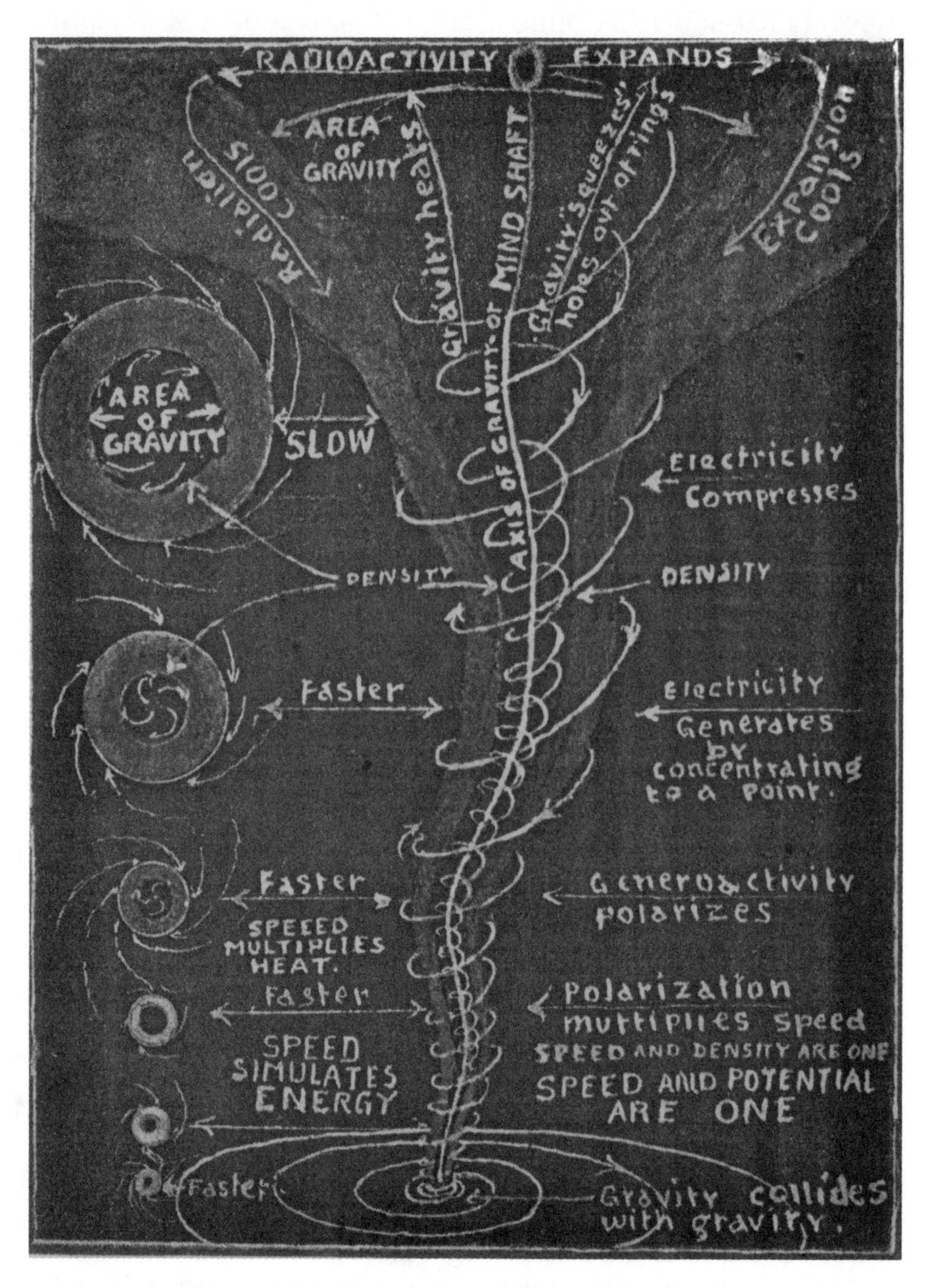

ELECTRIC COMPRESSION PRINCIPLES. Analysis of the relation of stillness to motion, and of energy to the expression of that energy.

By Dr. Walter Russell

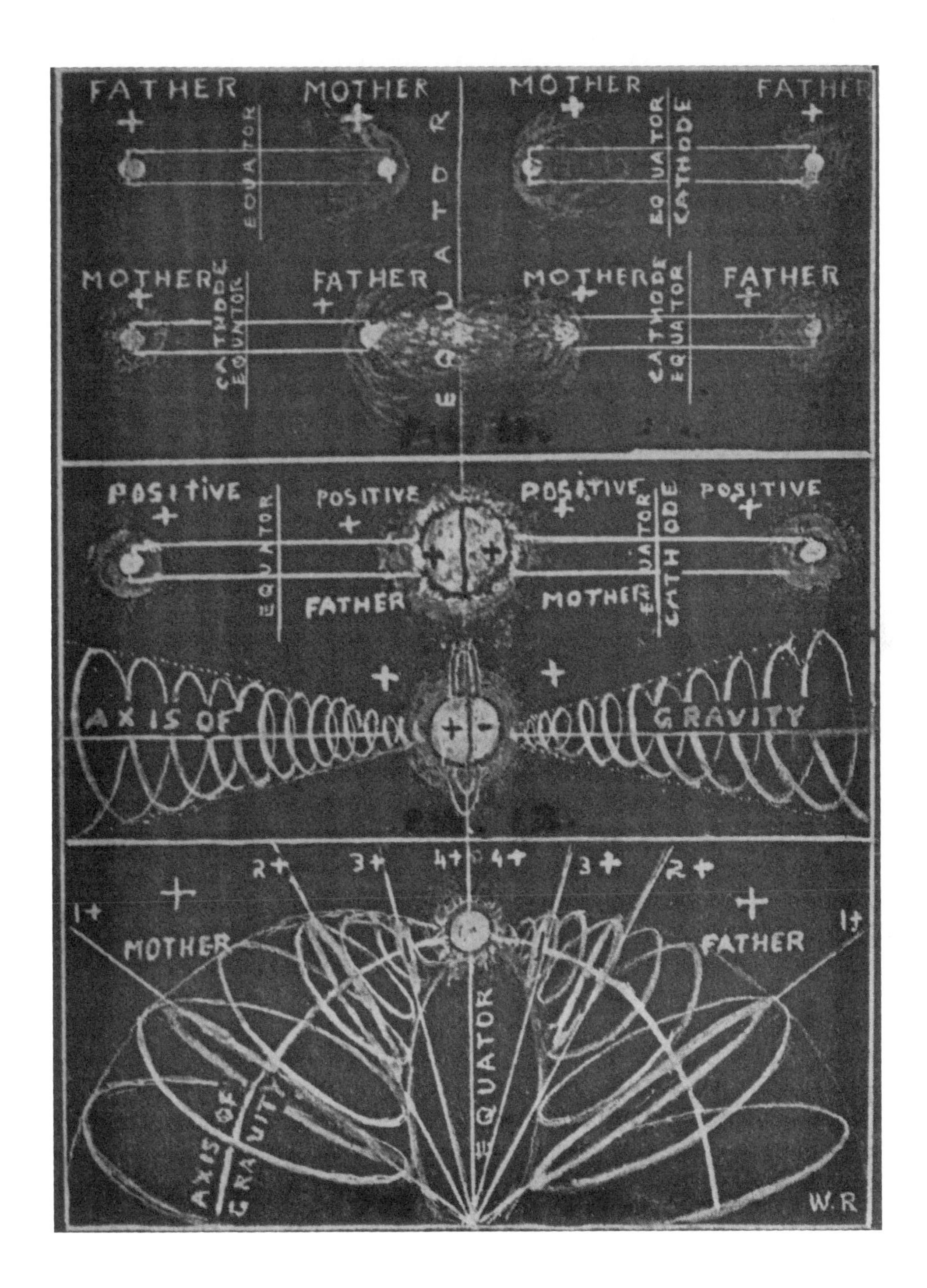

Illustrating the Father - Mother Principle of building bodies by dividing light into polarized units, and reproducing bodies by uniting two oppositely projected units into one by centripetal compression. POLARITY and sex are ONE. Sex and electric potential are ONE.

By Dr. Walter Russell

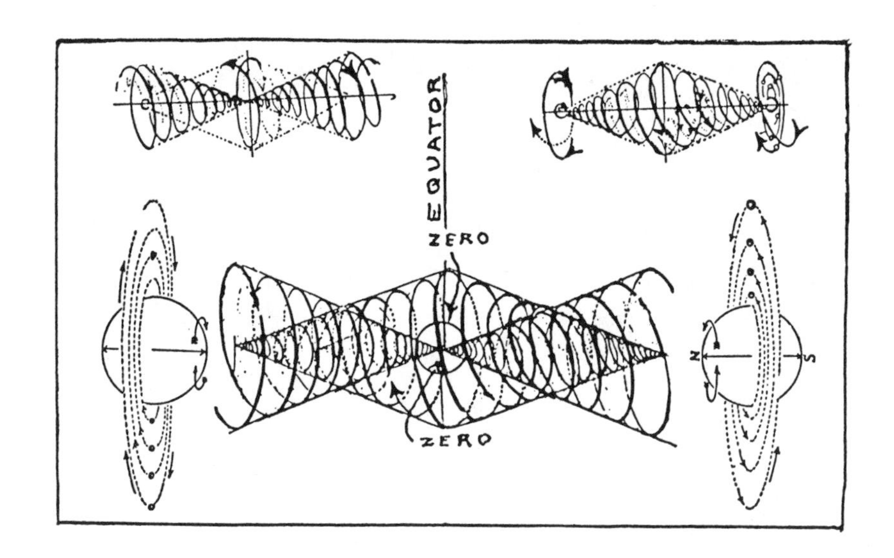

Complete life - death cycle as manifested in the electric current.

By Dr. Walter Russell

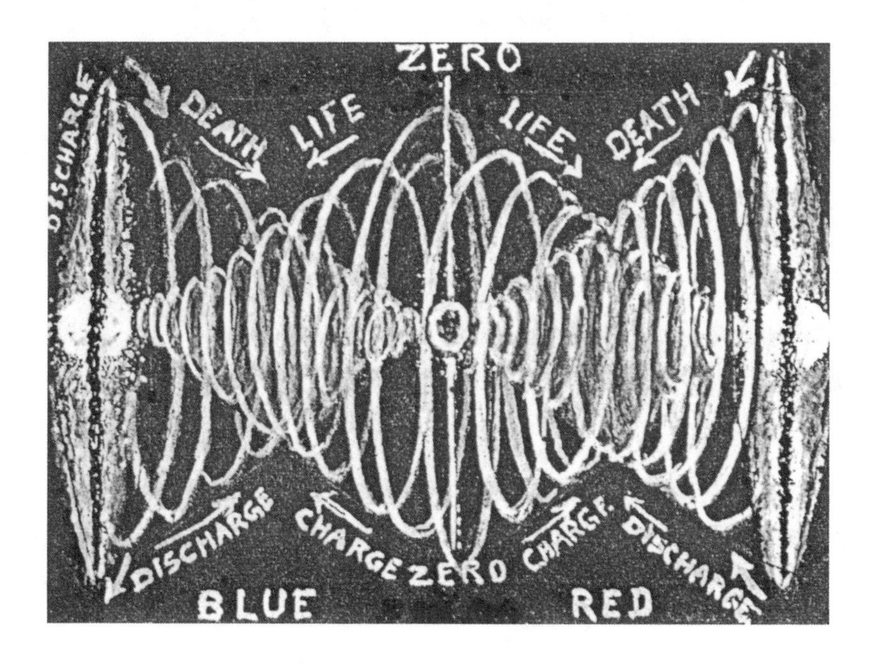

Complete life - death cycle as manifested in the heavens.

substance. *They will know it for what it is, which is motion only.*"

All that Russell is saying here is that energy arises from source and motion arises from energy, that matter as we know it is the product from the source. That the source is under great tension seeking to expand, while gravity is simply a kind of universal pump that creates focal points of compressed source we call matter.

Russell goes on to explain that vibration or oscillation is simply a swing from the outside of matter, to the center of gravity or mass, back to the outside again. Remember, gravity flows from the outside or void, to the center of motion or electric potential which is *rotary* by it's very nature. Vibration is the effect of the tensions created between the inside and the outside, the gravity axis and the motion around the axis.

Need I mention at this point that a vortex IS a gravity focal point? That the very motion of rotating or spinning indicates an exchange is taking place between the center and the outside of the *motion* whether we see it as a planet, sun, galaxy or the other direction, atoms. . .

As I understand, the center of our planet is *not* a dense molten core of iron that is magnetic and pulls us to it, but rather the center is a massless gravity focal point followed by gases then heat, immense heat. Very similar I assume, to a small sun.

From this point of view I understand that volcanos are not vents of lava being pushed up from the core of the Earth. Volcanos are only up and down in their structure from just under the surface, to the cone tube on the surface. Lava tubes run parallel to the surface and curve with the surface of our Earth.

Back to Russell. He states that the term "space" is misleading. And well it is.

When a wheel turns it must turn around something that is centering it. Likewise a planetary sphere must have a centering factor. Remember the toy top we cause to spin by pumping it down through the axis. The axis is both the channel for the energy and it centers the rotation created *by* the energy. The gravity wave creates an electric focal potential that *it becomes the center of,* as the potential starts spinning. The rest is simple. If enough gravity flows in, the electric focus now becomes a ring of energy that can't collapse to its center. It is fundamental in our teaching of electricity in school. When a current is flowing through a wire a field expands out from the wire. When the current is shut off, the field collapses. In Electronics 101 this is called "Back EMF" or back electro-motive force. If we prevent the electric ring focus from collapsing and keep the gravity pump going, lo and behold the concentration becomes mass to our five senses.

Personally I have wondered, thought, and researched about all of this for many years. The understanding doesn't come overnight except to a few, and that kind of Cosmic Illumination makes sitting in front of the television after work seem obsurd. When I look back to my "boy" years and remember telling my fourth grade teacher that atoms and solar systems run on the same principle, I smile a little. Also I am still understanding.

Russell Clarifies

Russell clarifies action and reaction. He writes: "Perhaps the most fundamental of misconceptions is the Coulomb electric law which says that opposites attract, and that gravitation also is a force which pulls inward from within, and that it attracts other bodies, when, in fact, both of these beliefs are just the opposite from the facts of Nature upon which they were misconceived. Coupled with this unnatural conception is the equally unnatural one to the effect that the universe WAS CREATED about two billion years ago, and is now radiating its energy away, instead of it BEING CREATED eternally."

He continues: "With that concept there is no room for an *uphill flow of energy,* there is but a downhill flow.

Knowledge of the nature of electricity would quickly dispel that idea that the universe itself is on its way to death. *There are two opposed actions* to every electric pulsation. One of them is GENEROACTIVE, which multiplies compression. That is Natures *uphill flow* which charges. The other is RADIOACTIVE, which multiplies expansion, and that is its *downhill flow,* which discharges. For this reason it is time that we begin to know the true nature of electricity and magnetism, rather than theorize from what our senses tell us."

So gravity is the generoactive pump of the universe. The electric motion is the gravity focal point. Natures way of storing energy manifests as a rotating electric field we know as an electrical potential - ring structure. Now what do we do with it? Let's consider the possibilities based on some established facts, and turn our attention to practical application.

Building Our Own

An author named RHO SIGMA has published some gravity research conducted by a man named John R.R. Searl. The book, "ETHER-TECHNOLOGY — A rational approach to gravity control" (available through Adventures Unlimited, see catalogue in back) contains some very interesting data and comments lending credence not only to the Ether Concept, but if indeed *form follows function* then one of the greatest mysteries of what "flying saucers" are and how they dart about so easily, is about to be unveiled.

Before we learn about Mr. Searl, you will be interested to know that gravity research and the actual flying of disc prototypes was being conducted in Germany as early as 1940. Some of these scientists had developed disc type craft that climbed to an altitude of 12,400 meters in three minutes and reached a top speed of 2,200 kilometers per hour. The craft could hover motionless and could fly as fast backward as forward. Some of the larger discs were as much as 50 meters in diameter.

An Austrian, Viktor Schauberger discovered what he called an "implosion" principle and invented a motor to take advantage of this new principle. The motor consumed only air and water while it generated light, heat and motion. In the implosion motor Schauberger said that diamagnetism was developed which made lift possible through "diamagnetic levitation". This in their day was the anti- gravity term we use so casually today. In Vienna, one ten foot model took off vertically at such surprising speed that it shot through the 24 foot high hangar ceiling and was blown to bits. Mr. Schauberger himself became the "property" of the U.S. government after the war, while several other gravity-disc researchers from Germany were taken to Russia for futher developments in the program.

When in the U.S. Viktor Schauberger alluded to the fact that his implosion motor operated on a water vortex principle where water was pumped around in a ring tube at the edge, and down through the center of the disc through another tube at 90 degrees to the plane ring rotation. This caused according to Schauberger, an atomic low pressure zone which developes in seconds when either *air or water* is caused to move radially and axially under conditions of a *falling temperature gradient* .

Private American and Canadian interest groups offered Schauberger $3,500,000 to divulge the secret of his motor, however Viktor would not demonstrate even a model until an international provisional agreement was signed. He finally signed with an American group but died suddenly 4 days later in September of 1958 where shortly before this he had told friends of his, "I don't even own myself any longer."

What has happened since in both the U.S. and Russia concerning this technology has undoubtedly been kept quiet. Tremendous advances must have been made somewhere along the line and any unbiased reader must admit that the German flying discs ushered in a new age of flight travel in *this century.* We recall the ancient Indian manuscript, "Science of Aeronautics".

Also in the 30's and 40's were Thomas Townsend Brown's

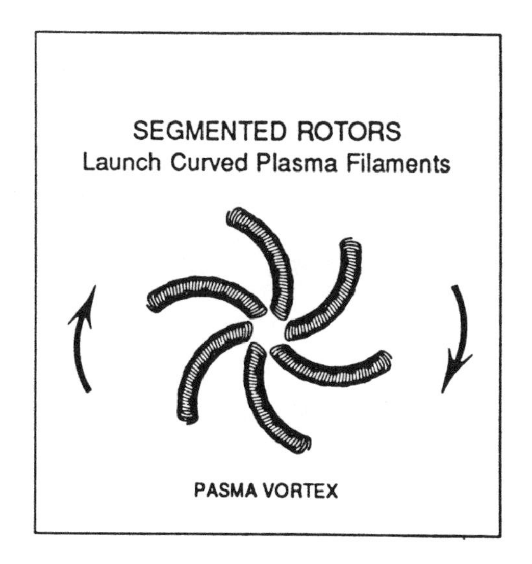

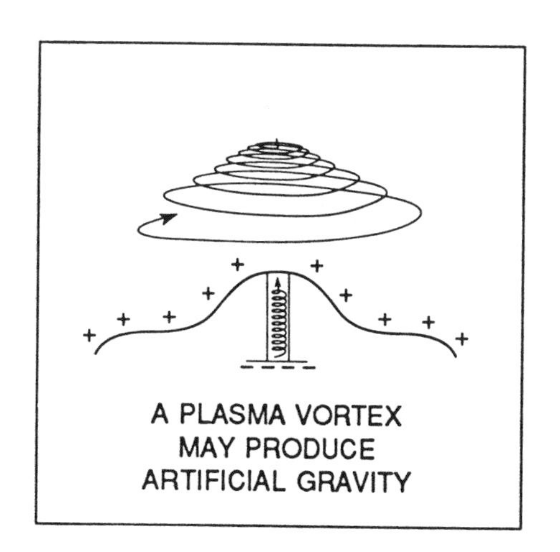

These are examples of vortex concepts presented by Moray B.King in his book " Tapping the Zero - Point Energy." Good reading for inventors.

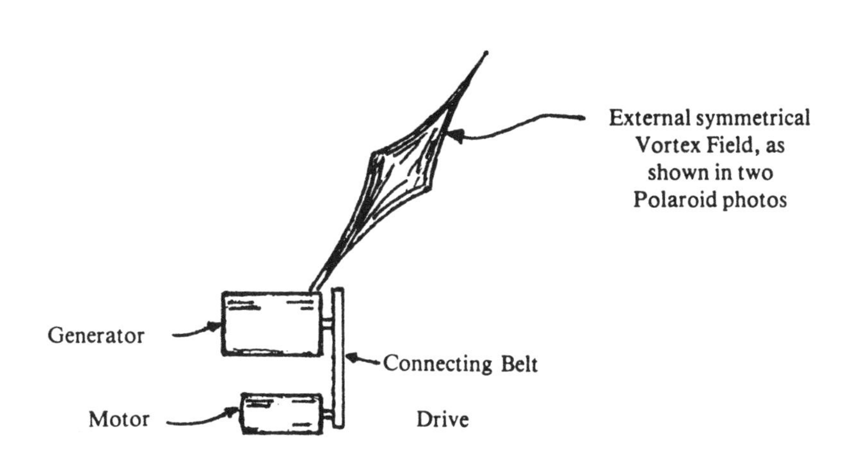

This is a drawing from a photograph taken in Canada, of a free - energy converter as it was being tested. What the camera recorded, appears to be energy extraction from space in the immediate vicinity.

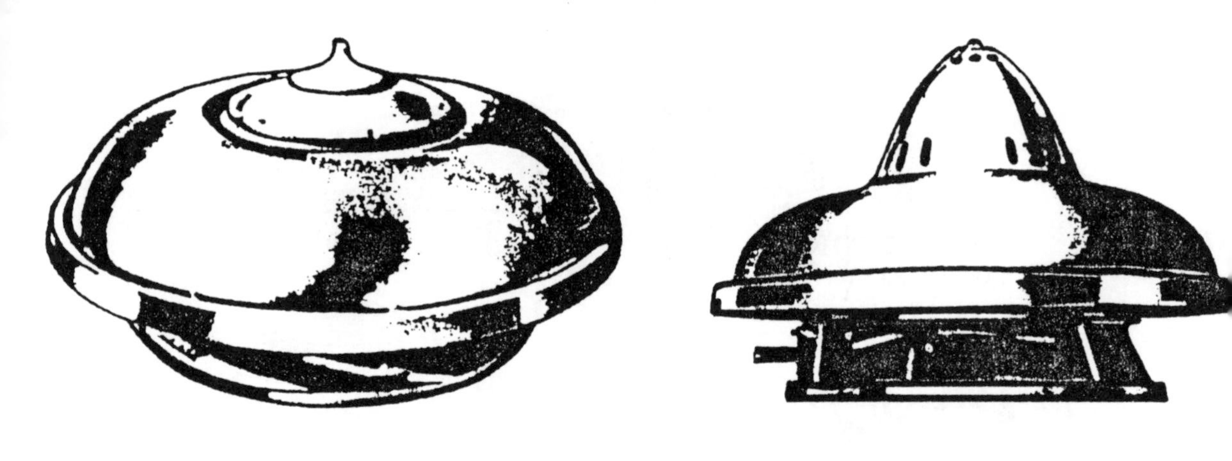

Above are two examples of Viktor Schaubergers "Implosion Motor" that caused what was referred to as diamagnetic levitation. Below are three variations of disc craft built by the Germans, based on Schauberger and team efforts.

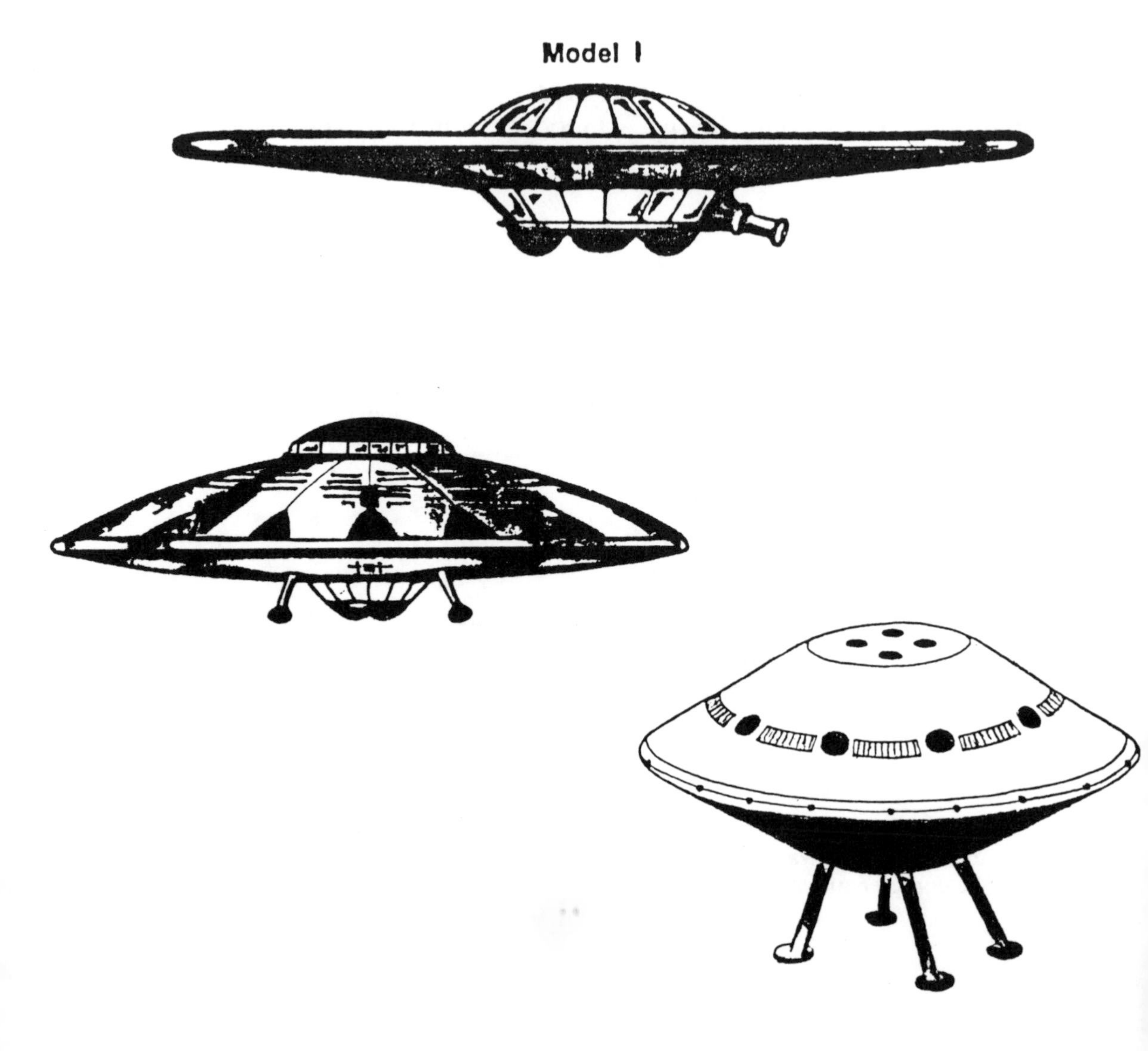

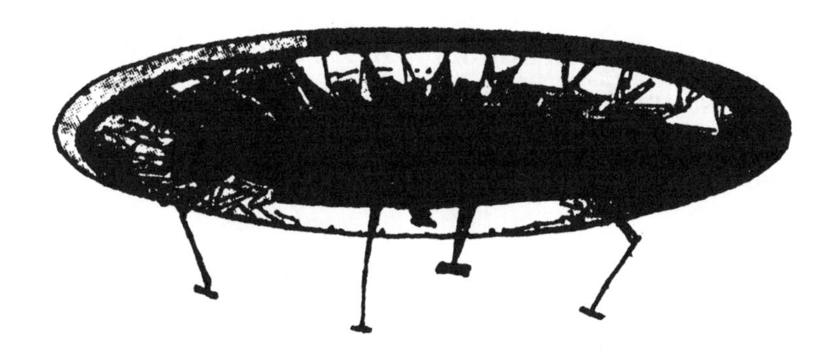

This, one of Searls flying prototype models was approximately 5 to 6 feet in diameter. The prototypes built later had a diameter of 30 feet.

WATER EFFECT UNDER DISC CRAFT U.F.O.
WALKER '89

Many separate incidents of this effect have been witnessed by people all over the world.

famous levity experiments where he demonstrated the tendancy of highly charged discs to lose weight and in some configurations to actually float. We recall earlier the science fair project of the young man.

Searl Ether Converter

John Searl was living in England in 1949 and employed at the time as an electrical fitter and cultivated much interest in electricity even though he had no formal education on the subject other than required by his job. Undaunted by conventional ideas he carried out his own investigations into electrical motors and generators. During work on this he noticed that a small EMF was produced by spinning metal parts whereby he proceeded to augment the effect by using slip rings in various ways. He noticed that when the annulus were spinning freely his hair bristled in the field created. He decided to build a generator based on the principle.

By 1952, the first generator (converter of ether) had been constructed and tested by Searl and a friend. It was about three feet in diameter and was set in motion by a small engine. The device produced the expected electrical power, but at an unexpectedly high potential on the order of 100,000 volts. While still speeding up the generator broke loose from the engine and rose to a height of about 50 feet. Here the rotor stayed for a while, *still speeding up,* with the air in the immediate vicinity of the rotor turning a pink like halo around it. Finally the whole works accelerated at a fantastic rate and is thought to have gone off into space. Since that day, Searl and others have made a number of small flying craft, some of which have also been lost. A form of control was developed and later craft 12 to 30 feet in diameter have been built.

Some interesting side light to the effects produced by the craft are:

1. Levity

2. Very high electrostatic fields.
3. The fields generated interact with radio receivers in the area.
4. Once the machine passes a threshold of potential, the energy output *exceeds* the energy input.
5. Above threshold potential, the generator (read: converter) becomes inertia free. It has *no apparent mass.*
6. The effects of the field around the craft that ionize the local air, also produce a *near vacuum* around it. However, the effects especially around the *equatorial plane* of the craft, have a tendency to push outside matter away from it. A type of force field as it were.
7. The prefered direction of travel at ultra-high speeds is away from the planet, the plane of the generator-rotor being at 90 degrees to the gravity field. When in horizontal flight the craft takes up an angle to the gravity field suggestive of the balance between two similar vector fields. In other words, the craft tilts or dips in the horizontal direction it is going.
8. Matter snatch during acceleration. This occurs when the craft is on the ground, and the drive is suddenly turned on. The rising craft takes part of the ground with it. If it were flying low over water, the water would peak up toward the bottom of the craft. The device cannot possibly be pushing itself away from the Earth if it is picking up matter from underneath.
9. If the craft hovers low to the ground too long, the ground becomes warm and the grass burnt. Also when flying in humid conditions precipitation or slight small cloud formations occur on the top side or around the craft.

The last effect, precipitation, was also noted by Nikola Tesla in his laboratory, for his energy producing coils were spiral configurations which created gravity focal points where water molecules became attracted above and around the top of the coil.

It should be pointed out with Searl's craft, that only a small amount of space fabric (ether) is converted for energy.

However, small changes in the ether lead to large physical effects because of the wave energy's high potential.

What I've listed above have been noted at one time or another as being effects from U.F.O. activity as well. The Searl Disc Ether Converter is real and is being worked on somewhere in the world. I can explain ALL of these effects simply by overlapping them with Gravity Vortex Mechanics.

9. — We'll start from the last to the first. I've covered precipitation. Now you'll need to use your mental holograms for this and picture an invisible upside - down tornado of whirling energy produced by the craft, and in fact it goes up *through* the craft with it's *focal apex above the craft.* When the device is low to the ground the wide end of the energy funnel is touching the ground. The Earth like all things, radiate the out-flow of gravity which is radiation. In producing the energy funnel low to the ground the device is also taking, from a confined area, some of the natural Earth Radiation that is being sucked up through the ground and grass in concentrated form. This of course produces the heating and burnt effect. If a Searl craft were to be positioned over a running automobile, I am certain the car would quit and all of the electrics would be dead until it passed over.

8. — Matter snatching during on ground acceleration is because of the funnel being turned on suddenly. You see, in a vortex, the wide end has a drawing effect toward it's apex. In fact, round neat holes have been cut into the ground by the Searl craft. U.F.O.'s have been known to fly over cars, picking them up and turning them around. Also since water is fluid, an upside-down waterspout takes the form of the gravity vortex underneath a craft over water. A craft on the ground is not sucked to the ground because the focal apex always being above the craft, easily overcomes any interaction below the craft equator.

7. — If a craft is flying straight up it is not tilting or dipping. However, by using directional coils within the craft, you can actually curve the vortex and thus the apex above moves off to one side. The *craft axis* now dips toward the

apex and the whole craft travels in that direction. The axis seeks the focus. The axis of a top in motion points to the gravity focus at the center of the Earth. The plane of rotation or equator of a top is at 90 degrees to the verticle gravity axis.

6. — Energy that has not been used to create the vortex, is expelled away from the craft at the edge or equator much like Saturn and her rings. This creates ionization of the air to form various glows or colors in the gases of the air. A kind of shell of energy, an actual field, is the effect. You can approach a craft from underneath if it is high enough, but you will find it difficult to approach directly from the side *unless* the field has been slowed or stopped. There is no room for air molecules between the field and the craft as a result. The benefit of a near vacuum side effect around the craft is that air now slips around the whole thing with no resistance from air friction.

5. & 4. — Above a threshold or where the inflow of ether has taken over naturally to produce the vortex, and all related fields, the craft now finds itself in the eye of the vortex. It finds itself in near stillness. All motion is taking place in a rotating fashion *around* the craft. An axis of apparent stillness is going up through the craft as the eye of a hurricane. Anything within the craft would not feel motion in the form of acceleration, sudden stops, sharp turns. In fact, you could turn off or dampen momentarily the focus in one direction, tilt your coils one-quarter the radius from where they were while re-establishing focus, and someone on the outside watching this would say you had just executed a 90 degree turn instantly.

Sometimes I feel that U.F.O.'s show off a bit. It's like they're saying; "Come on, figure this out."

3. — The associated fields through and around the craft have the most effect locally. However, secondary waves are produced which probably cover a wide spectrum including radio frequencies.

2. — We would expect very high electrostatic fields as a secondary effect. Very high electrostatic fields are produced in the presence of tornados as well.

1. — The craft in a manner of speaking, *falls in toward the*

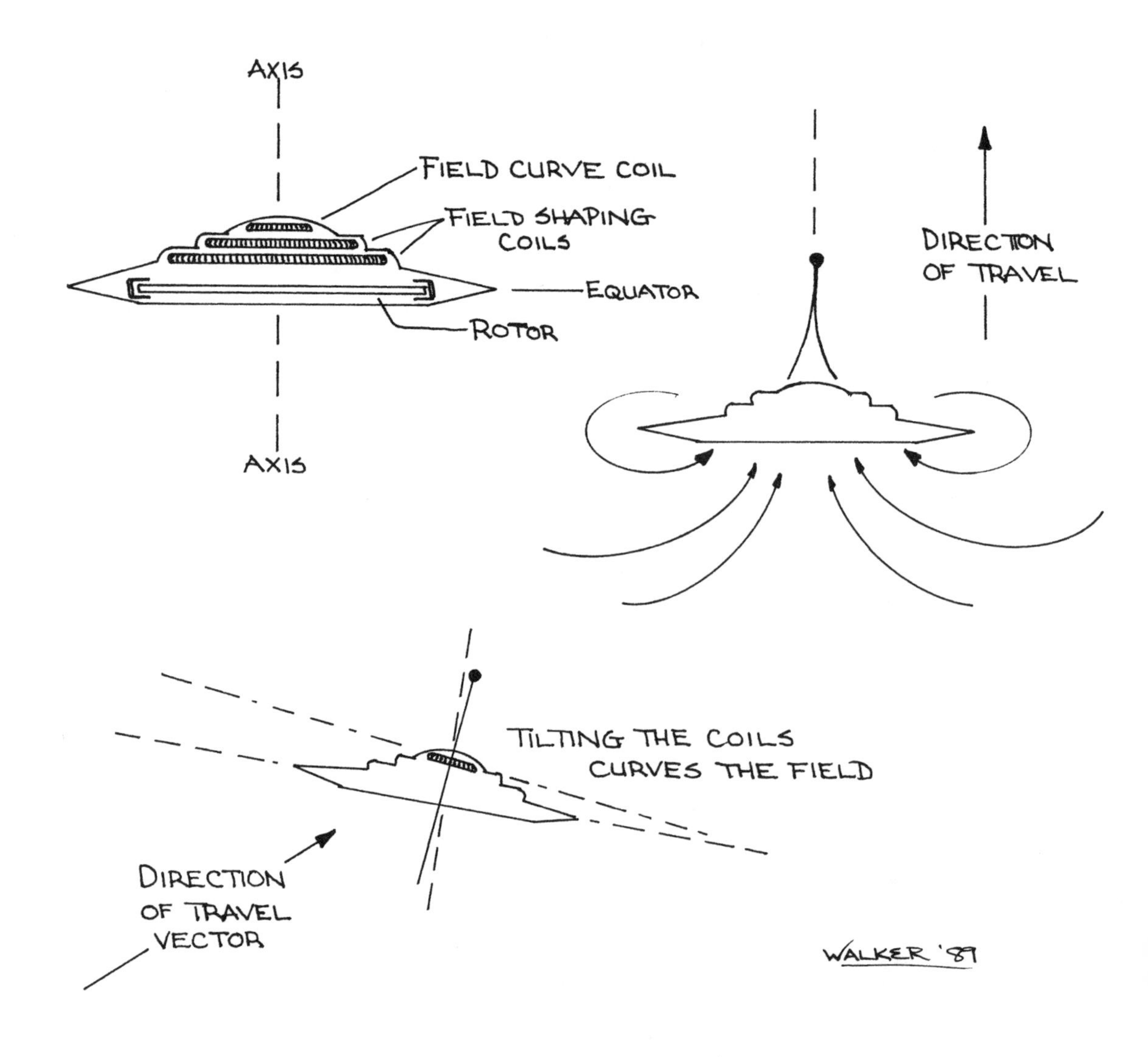

DISC DYNAMICS - Shaping and directing the vortex for travel by levity.

gravity focus above the craft. To say it is a controlled fall upward is a little confusing. Rather, the gravity field above the craft is more dense than local gravity on its way to the center of the Earth. It is more concentrated.

Form Follows Function

If I were to build a ring or rotor based motion system, the most natural type of vessel to contain it in would be round like a common motor. If I were to design my rotor like a flat plate, naturally my containment vessel would be saucer or convex lens shaped. The fact that it is lens shaped also assists the whole craft to act as a focussing instrument. If you've ever seen a magnifying lens used to make a fire, the smoke allows you to see light streams pass through the lens to create a cone. Where the cone is smallest at its apex, that's the energy concentration point.

Essentially the form of a saucer follows exactly its function to create a rotating potential - gravity vortex. It works with nature. The reason a disc craft would have more of a structure to it on the top side indicates room for pilot quarters and focal control coils, or in the case of remote craft, just the coils. I seriously doubt that a dome on top is for looking out of, then using it as an entry hatch as well. A jet fighter is one thing. A saucer, well...?

If you just build a flat rotor that produced a rotating field, space would curve in toward the center and then it would curve back out again. However, if you added a couple of more coils above the rotor, the top one being smallest, now you are shaping the field. A good example for a model would be a 3 foot rotor, a 2 foot coil above that, and your top coil is say - 9 inches in diameter. Now space curves in and is confined to smaller and smaller concentric rings. By the time it is leaving the topside it is almost pointed. It is focused. Focusing coils are not a new concept I just thought of. If you know even the basics of television you would know that focus coils are utilized to pinpoint and direct the tube electron beam against the screen. The principles speak for themselves.

Disclaimer

I have done a fair job in communicating certain principles which I myself consider to be universal in nature. By universal I mean they apply farther than we can see. I am not to be held responsible in the event that someone invents free energy and is not granted a patent, or in the event someone climbs on a gravity disc prototype, and was last seen clinging for dear life as it shot off into space.

I am also not responsible for furrowed brows, late night conversation, or excessively long trance-like states as a result of reading this material.

John T. Walker

John Walker has been a lifelong student, both on his own and in combined effort with other individuals, of the Universal Science or what are considered to be unestablished truths in science that are coming back around for a second chance in the minds of men. He has lectured for the U.S. Psychotronics Association, and has had contact with numerous authors/ researchers who have investigated controversal scientific issues.

Mr. Walker vanished in the Arizona desert in 1994, and hasn't been heard from since. It's thought that a dust-devil may have gotten him.

HOW I CONTROL GRAVITY
by
T. TOWNSEND BROWN

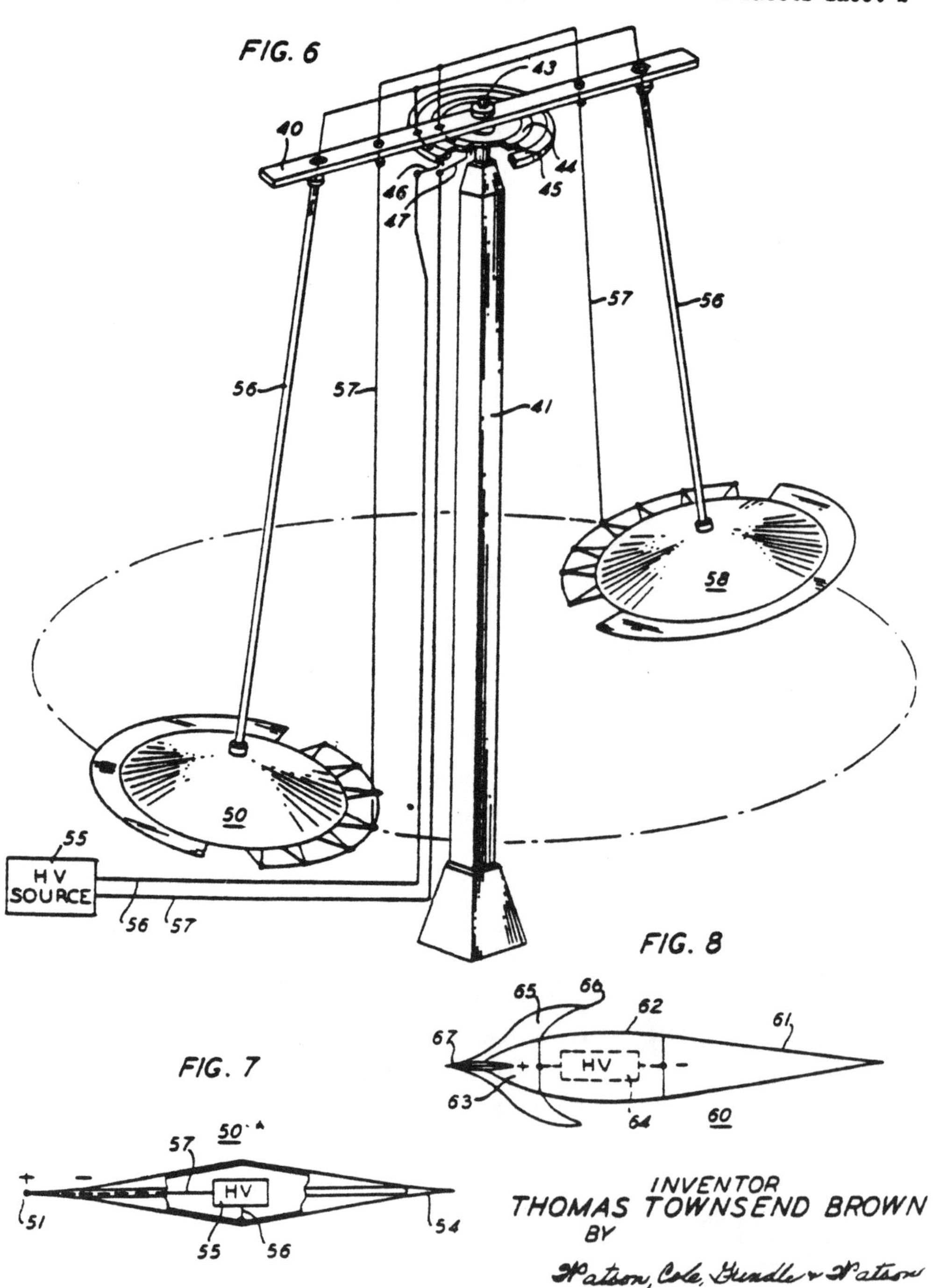

INVENTOR
THOMAS TOWNSEND BROWN
BY
Watson, Cole, Grindle & Watson
ATTORNEYS

An early article by Brown dealing, as the article preface said, *"with the meaning of the Einstein "Field Theories" and the relation between electro-dynamics and gravitation."*

There is a decided tendency in the physical sciences to unify the great basic laws and to relate by a single structure or mechanism such individual phenomenon as gravitation, electrodynamics and even matter itself. It is found that matter and electricity are very closely related in structure. In the final analysis matter loses its traditional individuality and becomes merely an "electrical condition." In fact, it might be said that the concrete body of the universe is nothing more than an assemblage of energy which, in itself, is quite intangible. Of course, it is self-evident that matter is connected with gravitation, and it follows logically that electricity is likewise connected. These relations exist in the realm of pure energy and consequently are very basic in nature. In all reality they constitute the true backbone of the universe. It is needless to say that the relations are not simple. Unfortunately at present they are not even well understood. The handicap is the outstanding lack of information on the real nature of gravitation.

The *Theory of Relativity* lends a new and revolutionary light to the subject by injecting a new conception of space and time. Gravitation thus becomes the natural outcome of so-called "distorted space." It loses its Newtonian interpretation as a tangible mechanical force and gains the rank of an "apparent" force, due merely to the condition of space itself.

Fields in space are produced by the presence of material bodies or electric charges. They are gravitational fields or electric fields, according to their causes. Apparently they have no connection, one with the other. This fact is substantiated by observations to the effect that electric fields can be shielded and annulled while gravitational fields are nearly perfectly penetrating. This dissimilarity has been the chief hardship to those who would compose a theory of combination.

It has required Dr. Einstein's own close study for a period of several years to achieve the results others have sought in vain and to announce with certainty the unitary field laws.

Einstein's new field theory is purely mathematical. It is not based on the results of any laboratory test and does not, so far as now known, predict any method by which an actual demonstration or proof may be made. The new theory accomplishes its purpose by "rounding out" the accepted Principles of Relativity so as to embrace electrical phenomena.

The Theory of Relativity thus supplemented represents the last word in mathematical physics. It is most certainly a theoretical structure of overpowering magnitude and importance. The thought involved is so far reaching that it may be many years before the work is fully appreciated and understood.

However, Dr. Einstein's announcement of his recent work has spirited the physicists of the entire world to locate and demonstrate, if possible, any structural relationship between electro-dynamics and gravitation. It is not that they questioned or doubted Einstein's reasoning or his mathematics (for they have learned better), but that they realized that the relation should exist and were eager to find it.

Early Investigations

The writer and his colleagues anticipated the present situation even as early as 1923, and began at that time to construct the necessary theoretical bridge between the two then separate phenomena, electricity and gravitation. The work was slow at first, due to the scarcity of information and lack of proper equipment, but the search gained speed as it gained information.

The first actual demonstration was made in 1924. Observations were made of the individual and combined motions of two heavy lead balls which were suspended by wires 45 cm. apart. The balls were given opposite electrostatic charges and the charges were maintained. Sensitive optical methods were employed in measuring the movements, and as near as could be observed the balls appeared to behave according to the following law: "Any system of two bodies possesses a mutual and uni-directional force (generally in the line of the bodies) which is directly proportional to the product of the masses, directly pro- portional to the potential difference and inversely proportional to the square of the distance between them."

It will be noted that this law is merely a combination and slight variation of Newton's law of gravitation and Coulomb's law of electrostatic attraction. In the specific text of this law the movement is in the negative to positive direction.

THE PECULIAR RESULT IS THAT THE GRAVITATIONAL FIELD OF THE EARTH HAD NO APPARENT CONNECTION WITH THE EXPERIMENT. THE GRAVITATIONAL FACTORS ENTERED THROUGH THE CONSIDERATION OF THE MASS OF THE ELECTRIFIED BODIES.

The newly discovered force was quite obviously the resultant physical effect of electro-gravitational interaction. It represented the first actual evidence of the very basic relationship. The force was named "gravitator action" for want of a better term and the apparatus or system of masses was called a "gravitator."

Since the time of the first test the apparatus and the methods have been greatly improved and simplified. Cellular "gravitators" have taken the place of the large balls of lead. Rotating frames supporting two and four gravitators have made possible acceleration measurements. Molecular gravitators made of solid blocks of massive dielectric have given still greater efficiency. Rotors and pendulums operating under oil have eliminated atmospheric conditions as to pressure, temperature, and humidity. The disturbing effects of ionization, electron emission, and pure electro-statics have likewise been carefully analyzed and eliminated. Finally after many years of tedious work and with refinement of methods we succeeded in observing the gravitational variations produced by the moon and the sun and the much smaller variations produced by the different planets. It is a curious fact that the effects are most pronounced when the affecting body is in the alignment of the differently charged elements and least pronounced when it is at right angles.

Much of the credit for this research is due Dr. Paul Biefield, Director of Swasey Observatory. The writer is deeply indebted to him for his assistance and for his many valuable and timely suggestions.

Gravitator Action an Impulse

Let us take, for example, the case of a gravitator totally immersed in oil but suspended so as to act as a pendulum and swing along the line of its elements.

When the direct current with high voltage (75-300 kilovolts) is applied the gravitator swings up the arc until its propulsive force balances the force of the earth's gravity resolved to that point, then it stops, but it does not remain there. The pendulum then gradually returns to the vertical or starting position even while the potential is maintained. The pendulum swings only to one side of the vertical. Less than five seconds is required for the test pendulum to reach the maximum amplitude of the swing but from thirty to eighty seconds are required for it to return to zero.

The total time or the duration of the impulse varies with such cosmic conditions as the relative position and distance of the moon, sun and so forth. It is in no way affected by fluctuations in the supplied voltage and averages the same for every mass or material under test. The duration of the impulse is governed solely by the condition of the gravitational field. It is a value which is unaffected by changes in the experimental set-up, voltage applied or type of gravitator employed. Any number of different kinds of gravitators operating simultaneously on widely different voltages would reveal exactly the same impulse duration at any instant. Over an extended period of time all gravitators would show equal readings and equal variations in the duration of the impulse.

After the gravitator is once fully discharged, its impulse exhausted, the electrical potential must be removed for at least five minutes in order that it may recharge itself and regain its normal gravic condition. The effect is much like that of discharging and charging a storage battery, except that electricity is handled in a reverse manner. When the duration of the impulse is great the time required for complete recharge is likewise great. The times of discharge and recharge are always proportional. Technically speaking, the exo-gravic rate and endo-gravic rate are proportional to the gravic capacity.

Summing up the observations of the electro-gravic pendulum the following characteristics are noted:

APPLIED VOLTAGE determines only the amplitude of the swing.

APPLIED AMPERAGE is only sufficient to overcome leakage and maintain the required voltage through the losses in the dielectric. Thus the total load approximates only 37 ten-millionths of an ampere. It apparently has no other relation to the movement, at least from the present state of physics.

MASS of the dielectric is a factor in determining the total energy involved in the impulse. For a given amplitude an increase in mass is productive of an increase in the energy exhibited by the system (E=mg).

DURATION OF THE IMPULSE, with electrical conditions maintained, is independent of all of the foregoing factors. It is governed solely by external gravitational conditions, positions of the moon, sun, etc., and represents the total energy or summation of energy values or levels which are effective at that instant.

GRAVITATIONAL ENERGY LEVELS are observable as

the pendulum returns from the maximum deflection to the zero point or vertical position. The pendulum hesitates in its return movement on definite levels or steps. The relative position and influence of these steps vary continuously every minute of the day. One step or energy value corresponds in effect to each cosmic body that is influencing the electrified mass of the gravitator. By merely tracing a succession of values over a period of time a fairly intelligible record of the paths and the relative gravitational effects of the moon, sun, etc., may be obtained.

In general then, every material body possesses inherently within its substance separate and distinct energy levels corresponding to the gravitational influences of every other body. These levels are readily revealed as the electro-gravic impulse dies and as the total gravic content of the body is slowly released.

The gravitator, in all reality, is a very efficient electric motor. Unlike other forms of motors it does not in any way involve the principles of electro-magnetism, but instead it utilizes the newer principle of electro-gravitation. A simple gravitator has no moving parts but is apparently capable of moving itself from within itself. It is highly efficient for the reason that it uses no gears, shafts, propellers or wheels in creating its motive power. It has no internal mechanical resistance and no observable rise in temperature. Contrary to the popular belief that gravitational motors must necessarily be vertical-acting the gravitator, it is found, acts equally well in every conceivable direction.

While the gravitator is at present primarily a scientific instrument, perhaps even an astronomical instrument, it also is rapidly advancing to a position of commercial value. Multi-impulse gravitators weighing hundreds of tons may propel the ocean liners of the future. Smaller and more concentrated units may propel automobiles and even airplanes. Perhaps even the fantastic "space cars" and the promised visit to Mars may be the final outcome. Who can tell?

IS ARTIFICIAL GRAVITY POSSIBLE?

by
Moray B. King

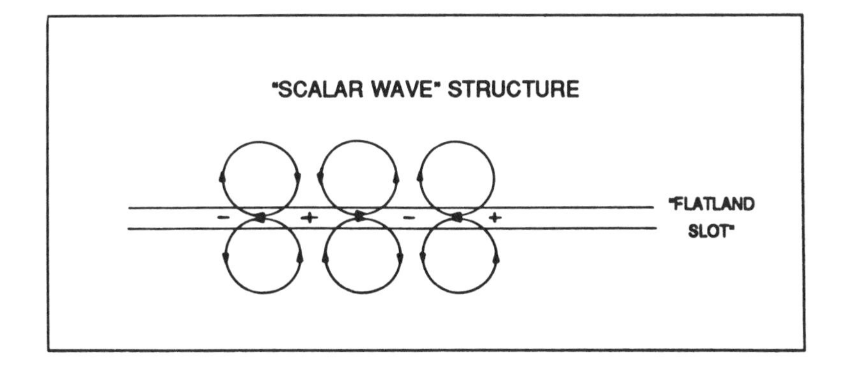

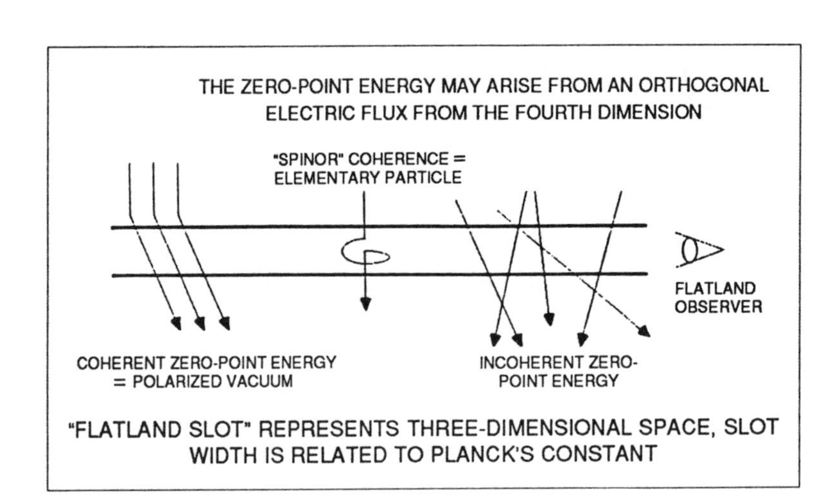
THE ZERO-POINT ENERGY MAY ARISE FROM AN ORTHOGONAL ELECTRIC FLUX FROM THE FOURTH DIMENSION
"SPINOR" COHERENCE = ELEMENTARY PARTICLE
FLATLAND OBSERVER
COHERENT ZERO-POINT ENERGY = POLARIZED VACUUM
INCOHERENT ZERO-POINT ENERGY
"FLATLAND SLOT" REPRESENTS THREE-DIMENSIONAL SPACE, SLOT WIDTH IS RELATED TO PLANCK'S CONSTANT

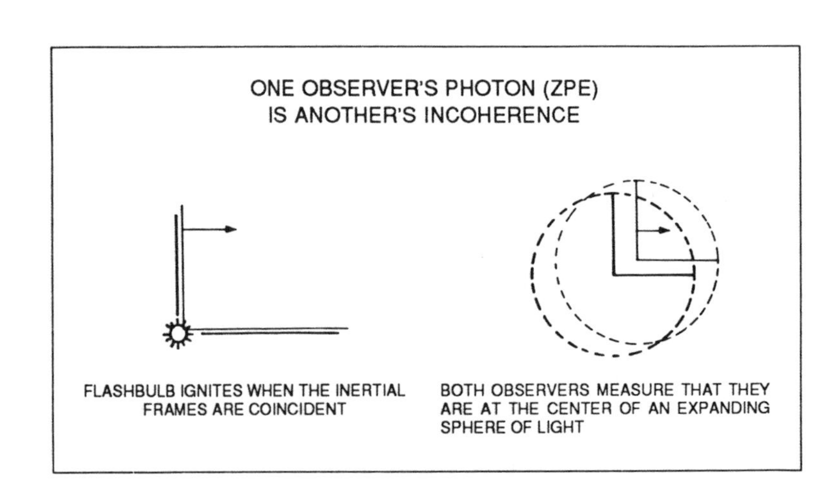
ONE OBSERVER'S PHOTON (ZPE) IS ANOTHER'S INCOHERENCE
FLASHBULB IGNITES WHEN THE INERTIAL FRAMES ARE COINCIDENT
BOTH OBSERVERS MEASURE THAT THEY ARE AT THE CENTER OF AN EXPANDING SPHERE OF LIGHT

Moray B. King, B.S. Electrical Engineering, M.S. Systems Engineering from the University of Pennsylvania, is currently a scientist at Eyring Research Inc., Provo, Utah. There he has co-developed an automated, broadband, antenna test system for field pattern testing of large HF antennas. As an advocation his main research interest involves supporting, with the standard physics literature, the speculation that a zero-point energy coherence can be induced by technological means. To encourage experimental research, he has given numerous presentations on this topic over the past 14 years.

ABSTRACT

Inducing a slight coherence in the action of the zero-point energy may curve the space-time metric yielding artificial gravity. The unidirectional thrust exhibited by stressed, charged dielectrics in the experiments of T. Townsend Brown may be evidence for this. A plasma vortex might enhance this effect for practical applications.

Is artificial gravity possible? If so, it would be an attractive means of propulsion because it would produce rapid acceleration without stress. According to general relativity, energy curves space-time producing gravity. If sufficient energy is placed above you, it can make you fall up. The mass equivalent of energy needed for levitation is about 10^{12} grams. If we have to provide all this energy, artificial gravity would be beyond today's technology.

However, modern quantum physics has within it an astonishing construction. It is the existence of the zero-point vacuum energy. Empty space is not empty. It consists of fluctuations of electricity whose energy density is on the order of 10^{94}grams/cm^3—an enormous number. This energy normally is unobserved because it self-cancels by destructive interference. However, if a device could induce just a slight coherence in the action of this energy over a region of space, then it could produce artificial gravity.

The work of T. Townsend Brown may give us a clue to how this might be done. A sufficiently charged capacitor can cause a vacuum polarization—a slight coherence of the vacuum fluctuations. Also the ionic lattice of a rapidly spinning body may interact with the vacuum energy causing a slight coherence which would alter the inertial properties of the body. This would occur because the vacuum energy itself curves space-time. Figure 1 illustrates the curvature of space-time with a two dimensional plane representing three-dimensional space. The lines are the path light takes as it travels through space. A massive body warps space, causing the path of light to bend. Energy also causes space to curve, as Einstein's theory of general relativity describes.

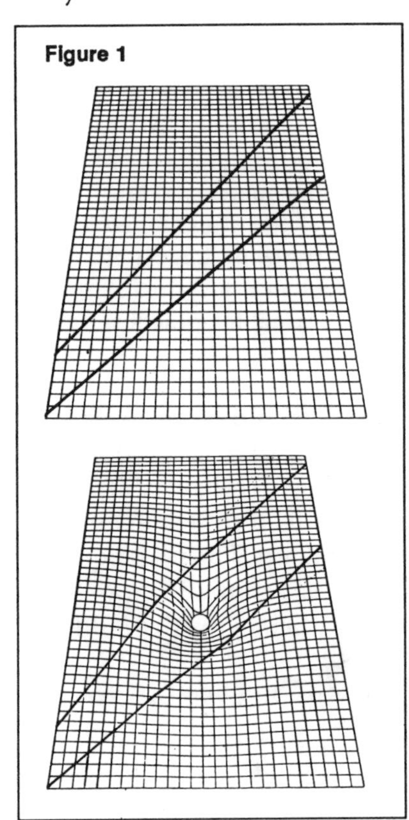
Figure 1

In Figure 2 the block diagram represents the ten nonlinear differential equations of general relativity. The box T represents the stress-energy tensor. It describes the location and flow of energy in space-time. The box g represents the metric. It de-

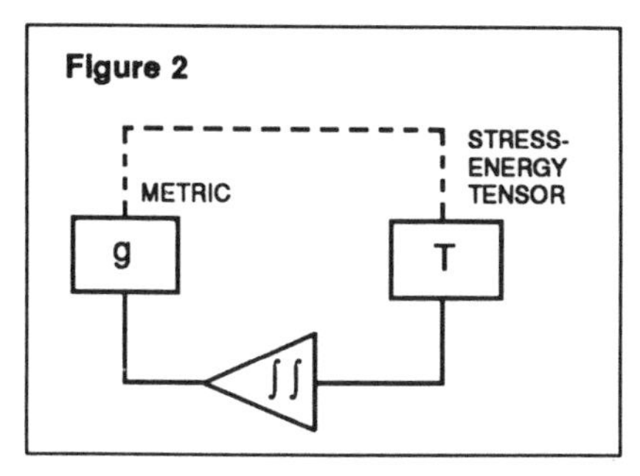

scribes the amount of space-time curvature that the stress-energy tensor induces. The double integration indicates that the stress-energy tensor controls second derivative terms of the metric. The dotted line represents an idea by Andrie Sakharov: The metric elasticity of space governs the action of the zero-point vacuum fluctuations, an energy which must be included in the stress-energy tensor. *This creates a feedback loop on a potentially active system.*

Can this system resonate? The nonlinearities of the system imply this may be possible. To illustrate, consider two boxes of energy and the curvature they induce at the point *P* (Figure 3). If we move the energy box *A* while holding the box *B* stationary, box B's contribution to the curvature at point *P* varies. If the system were linear, superposition would apply and B's contribution to the metric would be independent of A's. However, the system is nonlinear. For certain locations of A, B's contribution to the curvature is maximized. This idea also applies to a continuous field. As the field shape varies, the amount of curvature varies due to the mutual self-interaction of the field components. A resonant field is that field shape which maximizes the curvature of space-time. To efficiently curve space-time it is not only the amount of energy that is important, but *how this energy is used.* What is the coupling mechanism that allows mutual interaction of distant energy increments? It clearly has to be the fabric of space itself.

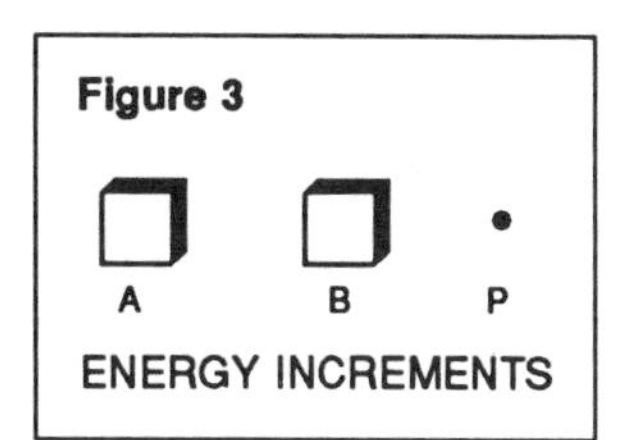

Microscopically viewed, space is a turbulent sea of energy consisting of electric flux (Figure 4). This flux enters from higher dimensional space through "mini-white holes" and leaves our three-dimensional space through "mini-black holes." To picture this concept, imagine our existence is confined to the two dimensional planar universe, flatland. We have no awareness of a third dimension. If a flux of energy were to pass through our space perpendicularly, we would have no awareness of this energy. However, if this flux jitters as it passes through flatland, a component of its motion would exist in our space. This flux component is the zero-point vacuum fluctuations. The diameter of these "mini holes" is on the order of Planck's length, 10^{-33}cm. The energy density through them is enormous, 10^{94}grams/cm^3.

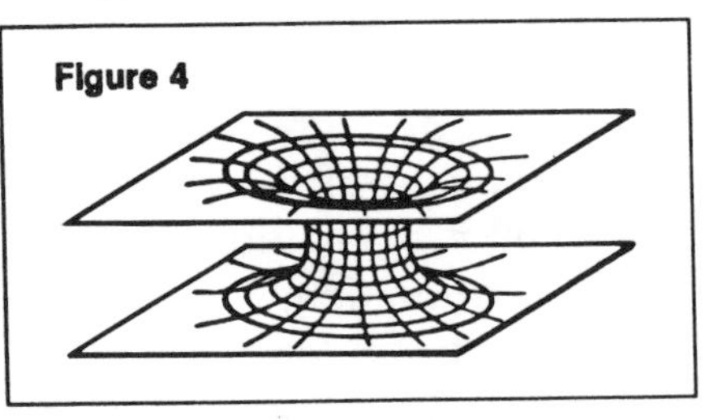
Figure 4

Large energy densities cause gravitational collapse. The top plane of Figure 4 represents three-dimensional space. The bottom plane also represents three-dimensional space —perhaps the same space. A large energy density causes space to pinch into what John Wheeler calls a *wormhole.* A wormhole can channel electric flux through higher dimensional space. It can connect distant points in the same three-dimensional space. In Figure 4, the plane represents three-dimensional space; the tube is the wormhole. Electric flux entering results in a "mini-white hole", flux exiting is a "mini-black hole." "Mini-holes" are constantly being created and annihilated in space, causing changing wormhole connections. Wheeler calls this resulting multiply-connected space *superspace.*

Can the vacuum fluctuations cohere in a region of space? Unstable particles or resonances may result from the temporary local coherence of the vacuum fluctuations. This model implies the existence of a whole spectrum of very small subnuclear particles that our science has not yet detected. The charge of a particle depends on a predominant flux from one type of mini hole.

The stable particles may likewise be a coherent alignment

of these holes. This model begets an interesting interpretation of the electron cloud around the nucleus of an atom. The electron literally is a cloud of negatively biased vacuum energy that maintains itself by a cohering self-connectivity through wormholes. This interpretation may shed insight on the wave-particle duality of matter. It illustrates coherence of the vacuum fluctuations in the quantum world.

Can the vacuum energy cohere over a large region of space in the macroscopic world? Note that levitation needs only a slight coherence in a statistical sense since the 10^{12} grams required for levitation is so much smaller than the 10^{94}g/cm^3. How can we achieve macroscopic vacuum polarization? Perhaps the work of T. Townsend Brown gives a hint.

Basically, Brown discovered that a sufficiently charged capacitor exhibits unidirectional thrust in the direction of the positive plate, and some types of capacitors exhibit more thrust than others. A type that worked very well consisted of 10,000 layers of lead foil and insulator. A dielectric consisting of a mixture of lead oxide and resin also worked well. Experiments with other materials lead Brown to the conclusion that a more massive dielectric with a greater dielectric constant produces a greater thrust.

T. Townsend Brown realized that the air around the capacitor's positive plate could be ionized and the fringe field would accelerate these ions back toward the negative plate causing the capacitor to move. In fact, J. Frank King, a collegue of Brown, patented a vehicle propelled by this type of ion propulsion (Figure 5). The top ring (21) ejects a plasma, and the rings (14, 15, 16) produce a magnetic field synchronously timed to accelerate the plasma downward.

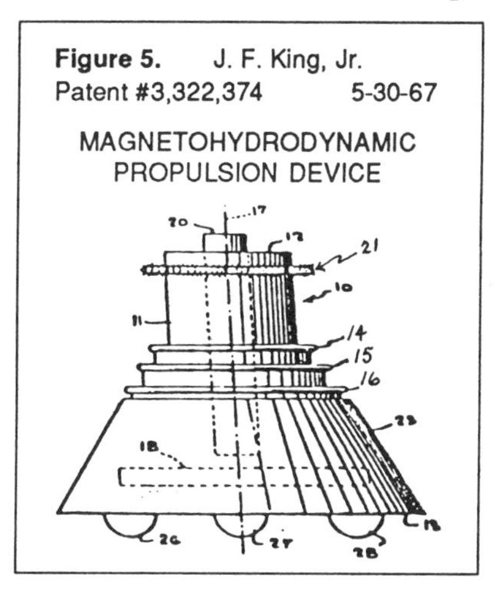

Figure 5. J. F. King, Jr. Patent #3,322,374 5-30-67

MAGNETOHYDRODYNAMIC PROPULSION DEVICE

The reaction force accelerates the vehicle upward.

To demonstrate that his capacitors involve something more than ion propulsion, T. Townsend Brown immersed them in oil, a medium that does not readily ionize. He observed that the thrust was almost the same as in air indicating that ion propulsion was not the major component of thrust. Brown charged the oil tank to the same potential as the positive plate in order to rule out electrostatic attraction as the cause of the thrust.

T. Townsend Brown also tested capacitors in a vacuum. He mounted two aluminum, open gap, parallel plate capacitors on a rotor. The vacuum pressure was monitored and held steady at 10^{-6} Torr. As he gradually increased the voltage from 90KV to 200KV, he observed irregular sparking concurrent with a large thrust. He also observed a residual thrust in the absence of sparking. The sparking occurred initially at about 15 second intervals. Its frequency gradually decayed until after about five minutes of operation, no more sparking occurred even though he left the rotor running days at a time. At 200KV the angular velocity would continue to increase, and he had to reduce the voltage to prevent the rotor from flying apart.

Running the rotor for days at a time that lead T. Townsend Brown to a remarkable observation. The capacitor thrust varied with the time of day even though the voltage, temperature, and pressure were held constant and carefully monitored. Over weeks of operation he observed a distinct sidereal correlation in the amount of thrust. This led Brown to believe that the charge capacitors were like catalysts that caused a vacuum polarization interaction with some type of energy flux hitting the earth from space. Perhaps the energy is from the sun; perhaps it is from the center of the galaxy. Brown is currently working with the Stanford Research Institute to determine the nature and source of this energy.

During the early 1940's T. Townsend Brown made a curious discovery. He found that enlarging and curving the positive electrode increased the thrust, and later he patented this concept (Figure 6). In the patent the large positive electrode is labeled (12); the negative electrode (14) and the dielectric rod connecting them (10). During World War II Brown discovered the optimum electrode shape. He described it as *triarcuate*—meaning "three arcs." He used a system of weights and pulleys to measure the thrust (Figure 7). When charged, a bright, colorful corona would appear on the surface of the triarcuate aluminum canopy.

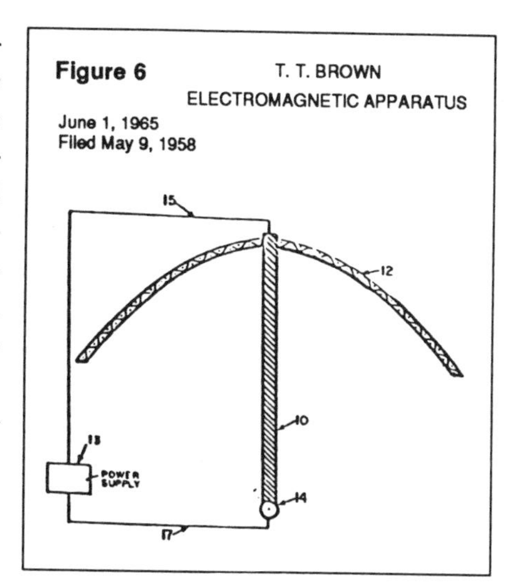

The factors that increased the thrust on the capacitor in T. Townsend Brown's experiments are:

1. Increase the plate area
2. Decrease the distance between the plates
3. Increase the dielectric permittivity
4. Increase the voltage
5. Increase the mass of the dielectric
6. Shape the positive plate

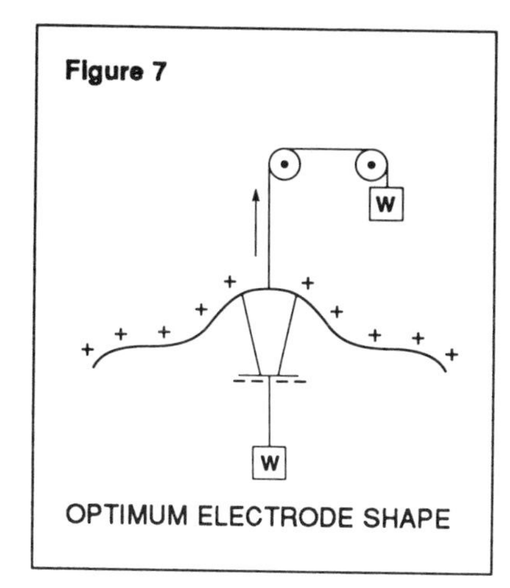

The first three factors increase the electrical capacitance of the apparatus. The thrust was approximately linear in voltage over the tested range of 50-300KV. Point 5 is what Brown feels links the nature of the capacitor thrust to gravity. Point 6 needs to be explained.

Any hypothesis that explains the thrust must include the

following key observations:

1. A large thrust was associated with a spark. A residual thrust existed without sparking. (In vacuum, 1956, gap capacitors).
2. A DC voltage (150KV) caused a thrust when initially applied. The thrust would decay within 60 seconds. Two minutes of recovery were needed at zero volts before thrust could be produced again. (In oil, 1928, dielectric of lead oxide and wax).
3. The thrust varied with the time of day. (In vacuum and in oil).

Point 1 requires identifying the source of the vacuum spark. Could it be due to air molecules that were trapped in the positive plate, due to electrons ejected from the negative plate, or due to both?

The second point was only observed in Brown's early capacitors of lead oxide and wax. It provides a clue to the optimum operating voltage for the capacitors. The dielectric should be polarized to the threshold of breakdown such that perturbations can cause avalanche breakdown. If the voltage is too high and the dielectric conducts, there will be no thrust. The thrust is associated with a change in state from polarization to breakdown. If the voltage is adjusted such that this change of state keeps repeating, a maximum thrust occurs.

Point 3 is a surprise and needs explanation.

Some possible hypotheses to explain the observations are listed below:

1. The surrounding medium is ionized and accelerated by the field. (Ion propulsion).
2. Avalanche breakdown through the dielectric is associated with:
   a. Plasma formation in the dielectric.
   b. An abrupt change in polarization.
   c. An abrupt change in the dielectric permittivity. A rapidly modulated dielectric permittivity may act as a transducer between electromagnetism and any

of the following:
1. Zero-point vacuum energy.
2. High frequency gravitational radiation.
3. High frequency permittivity waves.
4. Higher dimensional components of electromagnetism.
5. Neutrino flux
6. Ether flux

3. A resonant field is produced. The positive electrode is shaped to maximize the mutual interaction of the field with the metric. This may result in a spatially extended coherence of the vacuum energy fluctuations yielding a macroscopic metric fluctuation.

Ion propulsion is a component of thrust, but it cannot explain all the behavior. To illustrate, consider two equal size capacitors, the first with a small dielectric constant and the second with a massive material that has a large dielectric constant. Apply the same voltage to both. Brown observed that the capacitor with the larger dielectric constant exhibits the greater thrust. This is exactly opposite to what ion propulsion would predict since the fringe field of the first capacitor is greater. In the vacuum rotor experiment, only residual air ions accelerated in the fringe field can cause ion propulsion. The air ions in the main field will bang into the negative plate and impede the thrust. However, air ions in the main field between the plates can trigger breakdown—causing an electron cloud ejection from the negative plate.

The key to artificial gravity may be a rapidly accelerated, densely charged plasma cloud. It might cohere the vacuum fluctuations over a macroscopic region of space if there exists mutual coupling and connectivity of the particles in the cloud. Wheeler's superspace shows how a nonlocal connectivity may occur.

A simultaneous avalanche breakdown across the entire dielectric might macroscopically cohere the vacuum fluctuations in the region. If this were to occur in a perfect crystalline substance, the coherence could be significantly larger due to the regular ionic lattice's coupling to the vacuum energy. An outside energy flux could trigger and coherently couple the particles participating in the simultaneous breakdown.

What possible energy source could be interacting with the polarized field to account for the sidereal correlations? Could it be high frequency gravitational waves? Or could there exist such a thing as permittivity waves, perhaps generated by a plasma modulating a medium's dielectric constant? Brown ruled out standard electromagnetism by recent shielding experiments. But could there exist a higher dimensional form of electromagnetism that penetrates shielding? Could neutrinos be interacting? Could an ether flux exist that Michelson and Morley did not detect because the flux was perpendicular to the plane of their interferometer? (See references 5, 16, and 17.) These ideas are obviously speculative—only future experiments can give clues to develop a more concrete formulation.

Experiments by Bruce DePalma, N. A. Kozyrev and W. J. Hooper may provide a hint on how to magnify the effect. *Spin is the key.* Rapid spin of a plasma cloud or plasma toroid may result in a dynamic circular vacuum polarization. In a longitudinal magnetic field a plasma will naturally take the form of a spiral—a macroscopic helicon cloud. The optimum electrode structure can shape a helicon plasma into a vortex while guiding it into the fringe field. A plasma vortex may produce a macroscopic resonant field that slightly coheres the zero-point vacuum fluctuations to produce artificial gravity. (Figure 8) Also an inner, solid state plasma helicon through a dielectric or semiconductor may produce a strong vacuum energy coherence. The two plasma spirals may produce

Figure 8

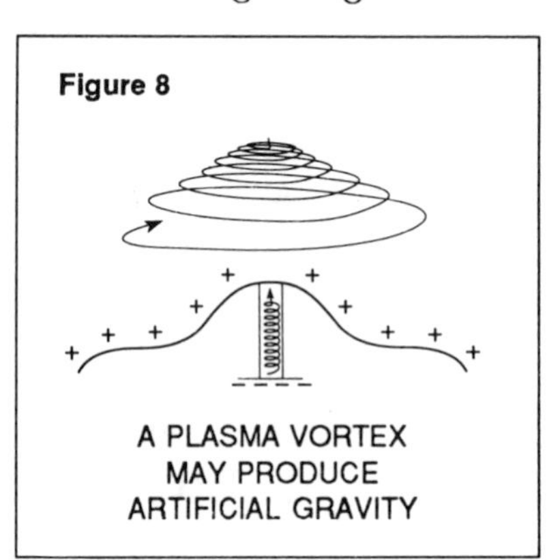

A PLASMA VORTEX MAY PRODUCE ARTIFICIAL GRAVITY

a toroidal vacuum polarization which may affect the inertia of neutral bodies in the region within. A pulsed ion vortex may produce artificial gravity sufficient for practical applications.

It is my hope that this discussion motivates others to continue the experimental investigations started by T. Townsend Brown, for a new propulsion technology awaits our discovery.

APPENDIX

On Coherence of the Vacuum Fluctuations
A Postulate of Physics

Artificial gravity relies on the coherence of the zero-point vacuum energy fluctuations. Many physicists believe that it is impossible to cohere the vacuum fluctuations because this would be a violation of the law of entropy. The law of entropy applies to those systems whose behavior is governed by a large set of *independently* acting components. Entropy is a statistical law stating that the chance alignment of random, independently acting elements is small. The probability approaches zero as the number of independent elements increases.

What is the true nature of the vacuum fluctuations? Are they independent "blinkers," or can an underlying connectivity occur, as described by Wheeler's superspace? The answer to this question is crucial in order to properly apply the law of entropy. In the case of random independent "blinkers," the law applies. The law of entropy may also apply to most cases in Wheeler's superspace formulation as long as the connectivity is random and nonlocal. However, if a device were to bias the connectivity in a region of space, the underlying assumption of *independence* would no longer apply, and it would be improper to invoke the law of entropy. *There is no proof in physics that states such connectivity is impossible.* On the contrary, there exist experiments that imply a vacuum energy coherence is occurring. (See references 7, 8, 9, 10, and 15.)

What is the true nature of the vacuum fluctuations? A postulate—that the vacuum fluctuations are random and independent—has divided modern physics into two camps. Most physicists today believe this postulate, and that it is improper to ask questions about the underlying causality. On the other hand, David Bohm, Jack Sarfatti and Fred Wolf postulate the existence of a possible underlying connectivity (as illustrated, for example, by Wheeler's geometrodynamics). This idea has been developing over the past twenty years and many physicists are unaware of it and its implications. The concepts are somewhat alien to classical physics, and difficult for many to understand since they invoke the existence of a physically real, higher dimensional space. For this reason, this postulate is currently not as popular.

But what governs physics—popularity or experimentation? There does exist a number of experimental anomalies[7-10,15] that can be explained by invoking the latter postulate that can't be explained in other ways. Most physicists have ignored these experiments, but this should not prevent capable individuals from repeating and verifying the work.

It is my hope that scientists will keep an open mind toward these investigations since recent theoretical developments in physics allow the possibility for an experimental success that could yield a tremendous technological advancement for mankind.

REFERENCES

1. Misner, Thorne, and Wheeler, *Gravitation,* W. H. Freeman, NY, 1970.
An excellent description of the zero-point vacuum energy fluctuations given in Chapter 43 and 44.
2. J. A. Wheeler, *Geometrodynamics,* Academic Press, Inc., 1962.
Geons and the vacuum fluctuations are described.
3. Toben, Sarfatti, and Wolf, *Space-Time and Beyond,* E. P. Dutton and Co., 1975.
A layman's introduction to multiply-connective space-time and vacuum energy is illustrated with numerous drawings.

4. R. Wald, "Gravitational Spin Interaction," *Physical Review* D, Vol. 6, No. 2, July 1972, p. 406.
The spinning body gravitational interaction is analyzed.

5. H. C. Dudley, "Is There an Ether?", *Industrial Research,* November 15, 1974; also in *Science Digest,* May 15, 1975, p. 57.
A neutrino flux model for the ether is presented.
See also *Il Nuovo Cimento,* Vol. 4B, No. 1, 68, (1971).

6. P. Bandyopadhyay and P. R. Chauduri, *Nuovo Cimento,* 38, 1912 (1965); 66A, 238 (1969).
Weak coupling of thé neutrino and the photon is discussed.

7. C. F. Brush, *Am. Phil. Soc.* V. 67, 105, (1928).
This paper describes an experiment which shows that aluminum silicate falls slower than other materials.

8. W. J. Hooper, *New Horizons in Electric, Magnetic and Gravitational Field Theory.* Electrodynamic Gravity, Inc., 543 Broad Boulevard, Cuyahoga Falls, OH 44221.
The author relates the motional electric field to gravity.

9. N. A. Kozyrev, "Possibility of Experimental Study of the Properties of Time," Sept. 1967, *JPRS* 45238, U.S. Department of Commerce, National Technical Information Service, Springfield, Virginia 222151.
An experiment relating the spin of mass to the pace of time is discussed.

10. B. E. DePalma, "A Simple Experimental Test for the Inertial Field of a Rotating Mechanical Object," *Journal of the British-American Scientific Research Association,* Vol. VI, No. II, June 1976.
The experiments are also described in the appendix of R. L. Dione, *Is God Supernatural?,* Bantam, NY, 1976.

11. C. C. Chiang, "On a Possible Repulsive Interaction in Universal Gravitation." *The Astrophysical Journal,* 1985, 87 (1973).

12. H. Bondi, "Negative Mass in General Relativity," *Rev. Mod. Phys.* 29, No. 3, 423, (1957).

13. J. A. Wheeler, "On the Nature of Quantum Geometrodynamics," *Ann. Phys.* 2, 604, (1957).

14. E. Streerwitz, *Phy. Rev.* D 11, No. 12, 3378, (1975).
The author's analysis includes vacuum fluctuations in the stress-energy tensor.

15. S. L. Adler, "Some Simple Vacuum Polarization Phenomenology..." *Phy. Rev.* D 10, No. 11, 3714, (1974).
16. J. Schwinger, "On Gauge Invariance and Vacuum Polarization," *Physical Review* 82, No. 5, 664, (1951).
17. Brill and Wheeler, "Interaction of Neutrinos in Gravitational Fields," *Rev. Mod. Phy.* 29, 465, (1957).
    This paper describes the interaction of neutrinos and gravitons.
18. P. A. Dirac, *Roy. Soc. Proc.* 126, 360 (1930).
    This paper first introduced the vacuum energy as an electron sea.
19. Gamow, *Thirty Years that Shook Physics,* Doubleday, NY, 1966.
    This text contains a layman's description of Dirac's Theory.
20. M. F. Hoyaux, *Solid State Plasmas,* (1970).
    This monograph is a concise introduction to solid state physics and plasma physics.
21. D. Bohm, "A Suggested Interpretation of the Quantum Theory in Terms of Hidden Variables," *Phys. Rev.* 85, 166, 180 (1952).
22. L. deBroglie, "The Reinterpretation of Wave Mechanics," *Foundation of Physics* 1, 1-5, (1970).
23. L. Motz, "Cosmology and the Structure of Elementary Particles," *Advances in the Astronautical Sciences,* V8, (1962).
24. H. Stapp, "S-Matrix Interpretation of Quantum Theory," *Phys. Rev.* D3, 1303, (1971).
    The S-Matrix "web" describes connectivity.
25. Hawkins and Ellis, *The Large Scale Structure of Space-Time,* Cambridge University Press, 1973.
    The basic physics of black holes is discussed.
28. D. Sciama, "Gravitational Waves and Mach's Principle," Preprint IC/73/94 from the *International Center for Theoretical Physics,* Trieste, Italy, 1974.
27. "Physics Made Simple," *Science News,* 106, 20 (July 1974).
    This article mentions experimental evidence that shows elementary particles are mini black holes.

FLYING SAUCERS:

SUPERCONDUCTING WHIRLS OF PLASMA

by
Hans Lauritzen

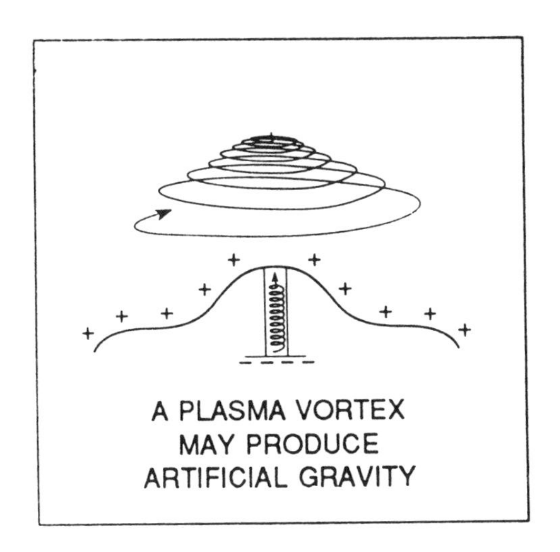
A PLASMA VORTEX
MAY PRODUCE
ARTIFICIAL GRAVITY

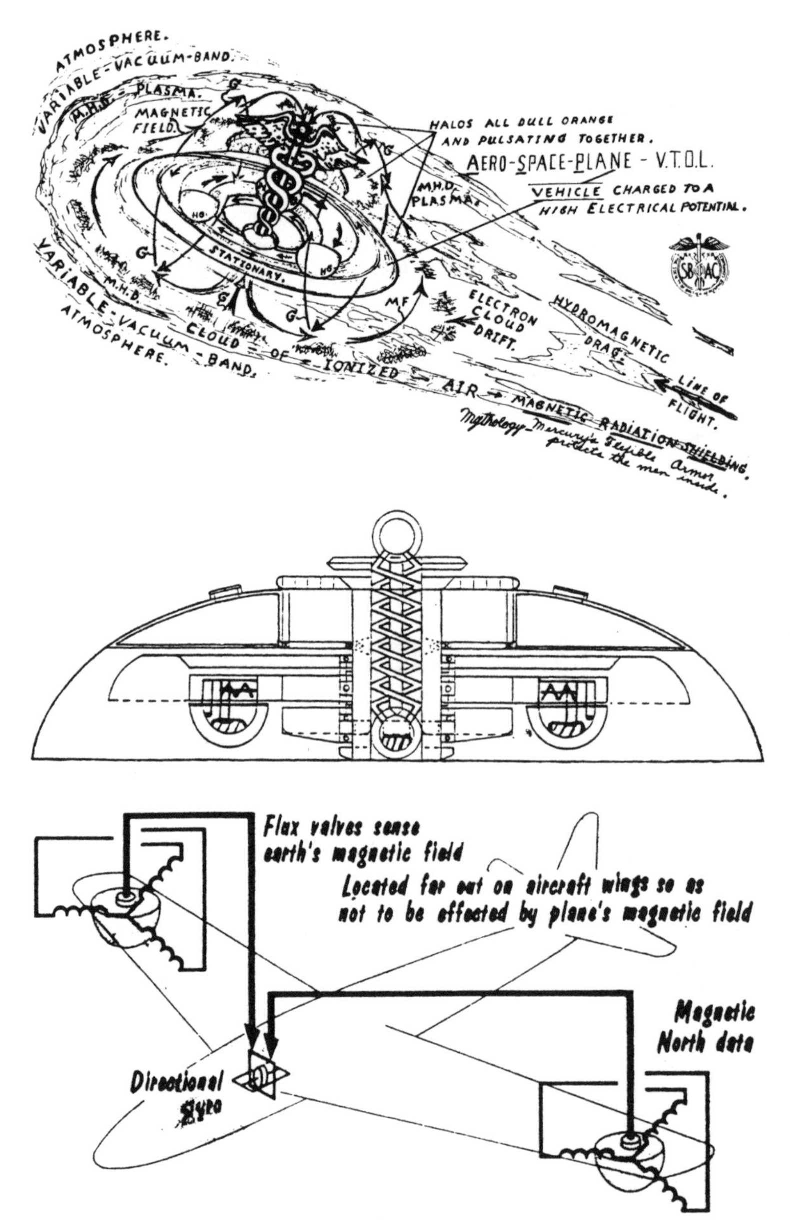

This page contains three diagrams from W.D. Clendenon's 1990 book, ***Mercury, UFO Messenger of the Gods.*** They demonstrate his *Mercury Proton Gyroscope* motor, which he feels is analagous to the ancient Cadeuseus symbol. The bottom diagram shows how the electrically charged mercury gyros operate in a similar to the flux valves of a directional gyro. The flux valves are used to keep directional gyros pointed to true north. They do this by acting as a synchro, using a particular set of voltages that is set up in a three stator winding sensing the earth's magnetic field. Curiously, they look like ping balls on the bottom side of the aircraft's wings, much like the half-domes on the underside of many discoid UFOs.

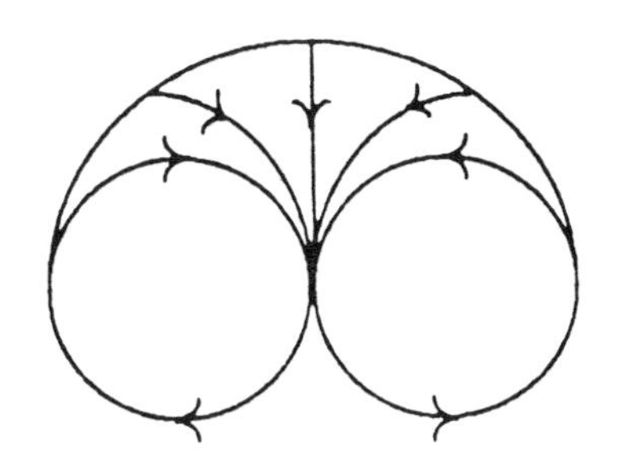

PLASMOID CURRENT FLOW

Reprinted from Flying Saucers magazine No. 51, March 1967.

In the following article you will find one of the best-kept secrets of modern times - and with its publication, it no longer remains a secret. This secret is the answer to the question most asked of the editors of FLYING SAUCERS: "What are the flying saucers?" Here is the answer, but only in part. The reader is warned that what he is about to read may be technical in its language, but the editors believe it is stated simply enough to be understandable to the readers of FLYING SAUCERS. To physicists (who are referred to the bibliography), the facts contained in this article will be obvious and unchallengable. In future issues, FLYING SAUCERS intends to continue this "definition" of flying saucers, and prove that the editors have been correct from the beginning in insisting the UFO is native to this planet, and not from interplanetary space. The editors invite scientists the world over to examine the facts presented here, to engage in further research, and to aid us in placing the riddle of the UFO where it belongs - in the realm of scientific reality, backed up by the scientific method. The editors of FLYING SAUCERS believe they are making history with this article, and with future enlargement of the demonstrable facts and theories contained therein. After nineteen years of mystery, a break-through has been made!

During recent years research on magnetohydrodynamics has provided humanity with a great number of discoveries of new physical phenomena. It has not been possible to predict these phenomena. We find ourselves at the beginning of a startling development, whose consequences are of immense importance for the future of humanity on planet Earth.

When a plasma is compressed by very strong magnetic fields, it has a tendency to form special shapes (plasmoids). The shapes assumed by plasmoids are unexpectedly rich and have provided information considerably in advance of the ability of theory to predict. A commonly observed hydromagnetic property of plasma is the tendency of the plasma spontaneously to form shreds or thin filaments. These filaments often form spirals in the container or magnetic bottles in which the plasma is accelerated.

When a plasma with high electrical potential is conpressed by a very strong magnetic field, and when at the same time a very strong magnetic field is created inside and along the axis of the column of plasma, the particles of the plasma will move in tight spirals around and along the magnetic lines of force. Associated with this corkscrew movement are two important qualities: the number of revolutions per plasma density is so high that the number of distant-encounter collisions per second exceeds the gyro-frequency, the corkcut takes place, so to speak, transforming the extremely tight spirals into a great number of ring-shaped plasma whirls. These ring-shaped plasma whirls are superconducting. As the electrical potential in them is enormous and now does not encounter any resistance, very powerful hydromagnetic shock waves are emitted.

For many years superconductors were thought of simply as substances whose electrical resistance vanishes below a certain temperature. This concept met with a number of minor but significant difficulties, which eventually led to the realization that superconductivity represents a state of matter in which a large number of electrons are in a coherent state of motion. Such a state can persist indefinitely in flagrant violation of the classical law that a circulating electron must lose energy by radiation.

A typical property of superconductors is their impenetrability for magnetism. As the ring-shaped whirls of plasma are super-conducting, magnetic fields cannot penetrate them, and they are actually imprisoned in their own magnetic fields (and those of the other neighboring whirls). That is why a ring-shaped current can remain stable indefinitely and at the same time emit powerful hydrodynamic shock waves without a supply of energy. These ring currents of superconducting plasma present certain analogies with the ring currents created in superfluid helium (below 2.2 degrees K) when bombarded with alpha particles, and with smoke rings.

Because of these dramatic macroscopic quantum effects exhibited by the superconducting ring currents, they can under certain conditions pave their way through the containers and escape from the laboratories without interruption of their coherent state of motion. They will then follow magnetic lines of force.

In a separate paper I have mentioned that such ring currents are created in the magnetosphere due to the interaction of ionized particle waves from the sun and the geomagnetic field. Earth's magnetic lines of force trap the waves of charged particles and cause them to move in spirals. Under certain conditions the spirals become so tight that huge stable superconducting ring currents are created. When the superconducting ring currents consist of trapped low-energy protons, they cause electromagnetic radiation at frequencies of about one cycle per second. And ring currents of trapped electrons give frequencies in the audible range, which is from 20 cycles to 15,000 cycles per second. The luminous phenomena are also called "HASER" an acronym standing for Hydromagnetic Amplification by Stimulated E-mission of Radiation. It is to be remembered that the particles are in a coherent state of motion in the ring currents.

A great number of ring currents are often combined in huge luminous cigar-shaped fields, but they can also separate and appear as smaller oval shapes or quite small luminous balls. Due to perturbations of the magnetic lines of force they can move nearer to the surface of Earth. Such strange bright lights (Unidentified Flying Objects) have been seen by millions of people. The shapes reported correspond very well to the shapes of superconducting ring currents. As they are very sensitive to magnetic lines of force, they have often been reported hovering over or moving along powerful electric current wires. When following the magnetic lines of force of Earth, the path will look like a zigzag line. They will also tend to move along certain geographic lines because of deviations of the geomagnetic field due to deposits of magnetic material in the Earth (orthotheny).

UFO s are often changing color when changing speed. In the superconducting ring currents the light is emitted by electrons that exhibit coherent wave properties over macroscopic dimensions. And a magnetic field is capable of varying the energy of quantum levels and hence the frequency of the emitted light, as well as reducing drastically the threshold for the onset of laser action.

Many other strange phenomena accompany the UFO s. The strong hydromagnetism gives magneto-optic reflection and absorption, so that the ring currents can look solid and metallic. Also magneto-acoustic effects are common. In the superconducting ring currents the electrons are in a coherent state of motion thus producing coherent electromagnetic waves as well as hydrodynamic shock waves are capable of giving rise to compression and rarefaction of particles. The free electrons in a current will thus change their density. Now, the behavior of electrons in a metal follows the rules of Fermi statistics, according to which the energy of electrons is always related to their density. Thus when the density of the electrons in a metal is disturbed, the distribution of their energies is also disturbed. Electric currents will be stopped causing black-out of electric powered machines and instruments. To re-establish the equilibrium distribution a certain relaxation time is required. Then the electric powered devices will reassume their function as if nothing had happened. Such black-outs are some of the most reported effects during close approach of UFO s.

Two theories have been put forward in order to explain the enormous amounts of energies released by hydrodynamic plasma. The first one says that there are hidden or uncounted extra positive and negative charges and hidden or uncounted magnetic movement in the plasma. The second theory says that the energies are released when the hydromagnetic whirls of plasma assume the same geometric dimensions as the electrohydromagnetic threads appropriate for an elementary particle. The Mach principle taken into consideration, the self energy of any elementary particle must be described within the framework of the universe as a whole. The universe is considered as a transient stage in which a transient latent universal potential is transformed into quanticized active local energies by processes inside elementary particles. Now when we know that the energies are released during a superconducting state of plasma, the first theory can be ruled out definitely. The second theory is still valid. The superconducting state of electrohydromagnetic threads in particles and of coherent whirls of elementary particles present us only with the information on how infinite amounts of energies can be released, but it does not provide us with any information on why and from where the energies are coming. Any hypothesis saying that the superconductivity in itself possesses self-energy infinites must be considered as irrelevant. The theory on a transient latent universal potential has considerable conceptual merit in as much as it also conforms to cosmologic concepts and to universal constants.

The following literature is recommended:

Malcolm McChesney: "Shock Waves and High Temperatures", Scientific American, February 1963, (415 Madison Avenue, New York, N. Y. 10017, USA).

Henry H. Kolm and Arthus J. Freeman: "Intense Magnetic Fields", Scientific American, April 1965.

"Outer Atmosphere Emits Laser-Like Light", Science News, September 17, 1966, (Scientific Service, Inc., 1719 N St., N. W., Washington, D. C. 20036, USA).

Harold W. Lewis: "Ball Lightning", Scientific American, March 1963.

Richard D. Mattuck: "Supraledning", Ingeniøren-Forskning, 15. August 1966, Teknisk Forlag A/S, (Skelbaekgade 4, Copenhagen V, Denmark).

Frederick H. Mueller: "Magnetohydrodynamics", Selective Bibliography of 124 reports during the period of 1950 to September 1960, U. S. Department of Commerce, (National Bureau of Standards, Institute for Applied Technology, Clearinghouse for Federal Scientific and Technical Information, Springfield, Virginia, 22151, USA).

Luther H. Hodges: "Magnetohydrodynamics", Selective Bibliography of 185 reports during the period of August 1960 to January 1962, U. S. Department of Commerce.

"Plasma Study of Lockheed Aircraft Corporation", Science Newsletter, November 6, 1965, (Science Service, Inc.).

Klaus Dransfeld: "Kilomegacycle Ultrasonics", Scientific American, June 1963.

Pietro Banna: "Il Princilio di Scambio (Banna) e i Suci Rapporti con la Relativitá Finale di Fantappie-Arcidiacono", Centre Européen pour les Recherches sur la Gravitation, (Via Borghesano Lucches 24, Rome, Italy). Supplement au Cahier "G" n. 15, 1965.

Winston H. Bostick: "The Gravitationally-Stabilized Hydromagnetic Model of the Elementary Particle", Gravity Research Foundation, (New Boston, N. H. , 03070, USA).

ANTI-MASS GENERATORS IN UFO PROPULSION

by
Kenneth W. Behrendt

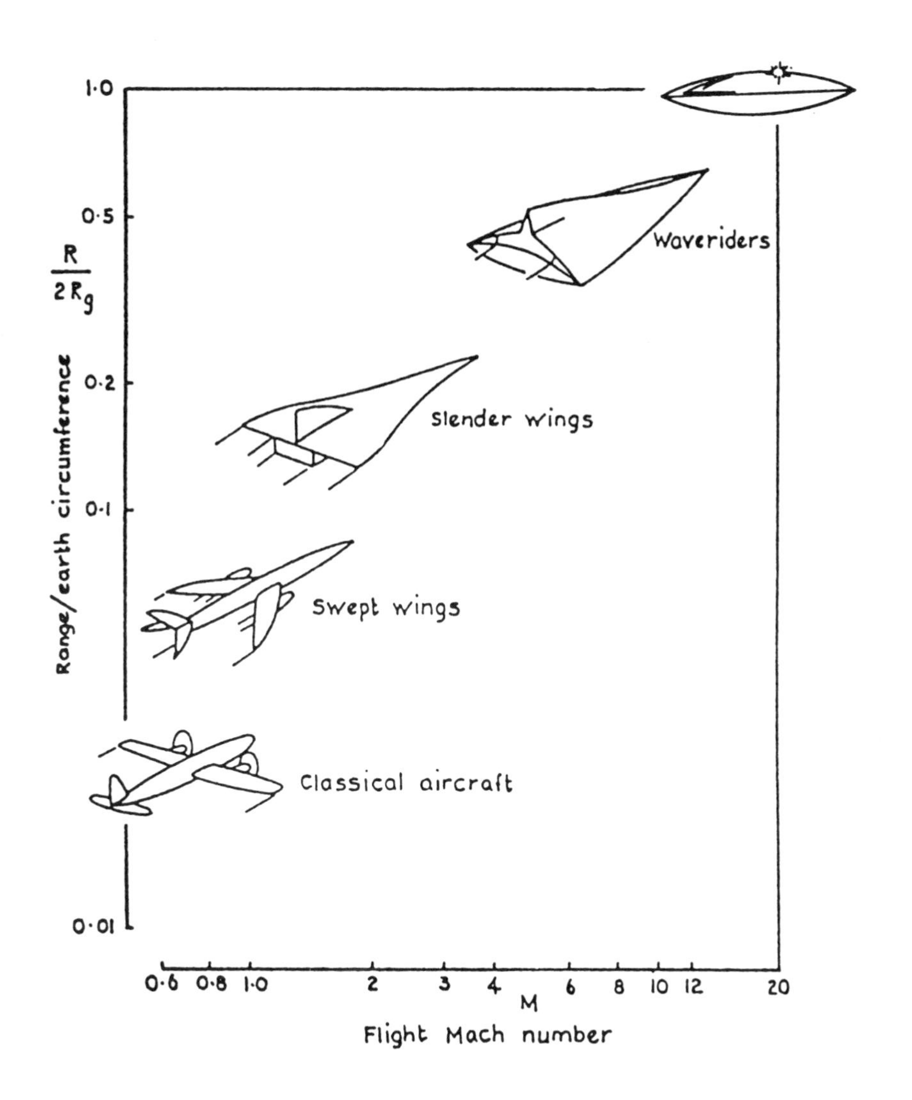

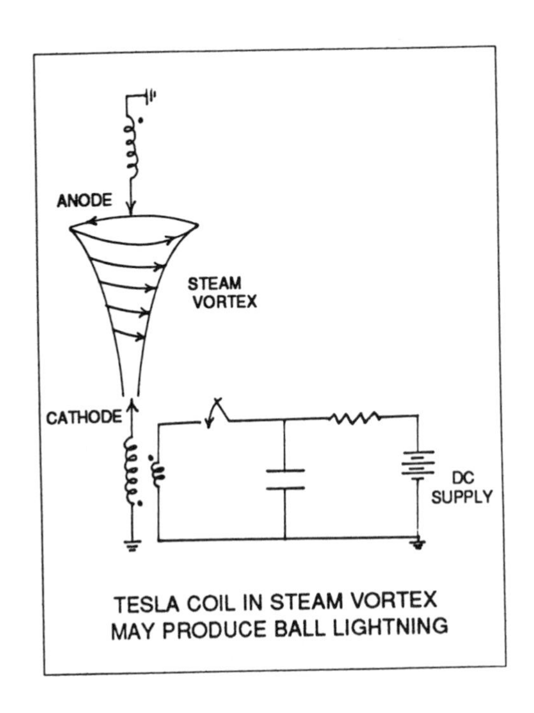

TESLA COIL IN STEAM VORTEX
MAY PRODUCE BALL LIGHTNING

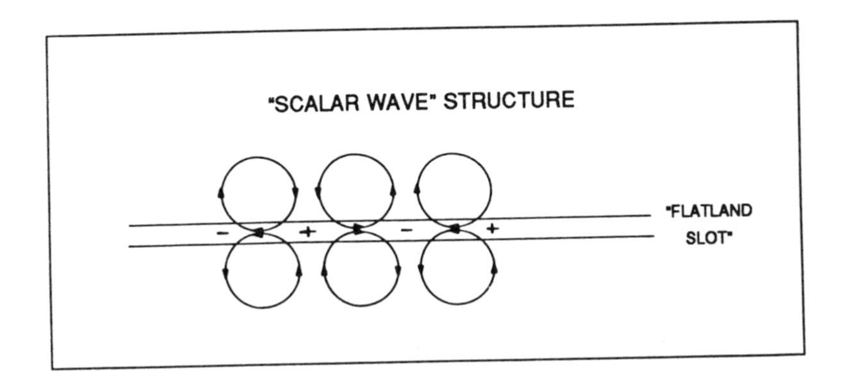

All materialistic ufologists hope for a satisfying solution to the so-called UFO mystery that will explain the origin of these craft, the purposes of their crews in visiting earth, and the technological means by which this is achieved. Unfortunately, in the absence of direct and reliable contact with extra-terrestrial beings, we may *never* know from what star systems our visitors originate or their purposes in earth's atmosphere and oceans. However, without direct contact we can still determine the technological means by which their vehicles operate. By carefully collecting accurate and detailed sighting data, organizing it and extracting the overall behavior and performance patterns it contains, the scientific method should allow us to determine, in time, the means of propulsion utilized in UFOs and how *all* of the various secondary effects are produced.

The final hypothesis that provides this partial solution must meet four criteria. First, it must adequately account for all of the data presently available on UFO structure and performance; secondly, it should be internally consistent, which simply means that one part of the hypothesis should not contradict another part of the same hypothesis; thirdly, this final hypothesis must be predictive so that it can explain UFO-related phenomena not originally used in the formulation of the hypothesis; and fourthly, the hypothesis must be detailed enough so that it will be capable of experimental verification or denial.

Unfortunately, in the last forty years ufology has concentrated mainly on the acquisition of UFO data and only a small percentage of the credible literature deals with attempts to rationalize the performance capabilities of UFOs and the secondary effects associated with their presence.

Practically all of the hypotheses advanced in an effort to provide the desired solution to the mystery have failed to meet the necessary criteria for good, workable hypotheses. Most often, they could not account for all of the available data and would become internally inconsistent when efforts were made to extend them. If the theorist lacked training in the contemporary physics of his day, his hypothesis might be so vague or qualitative that it could not be used to make predictions about the, as yet, not precisely known UFO phenomena, such as the structure and arrangement of a UFO's internal propulsion equipment. In almost every case, the hypotheses advanced lack the mathematical rigor needed to experimentally test their validity.

As can be seen, the attainment of a final hypothesis that does meet these difficult criteria is highly desirable because it should allow us to readily *duplicate* what has been so frequently demonstrated for us by our extraterrestrial visitors. With the duplication of the UFO phenomena we can expect an immediate and profound change in the way we live our earthly lives. Transportation will be revolutionized as we manufacture our own UFO-like craft that can easily fly us from one side of our planet to the other in minutes. We can finally complete the exploration of our solar system, establish colonies on some of the other planets or moons of our system, and begin planning our race's first interstellar voyages. There should also be tremendous advancements brought about in the fields of energy generation, medical science, metallurgy, physics, etc.

With the above in mind, let us now explore some of the attempts made by ufologists in the last forty years to shed some light on this all-important aspect of the subject. Although the literature relating to propulsion and secondary effects is sparse, it is still not possible to give a full and detailed treatment of all of it within the confines of a single chapter. However, an attempt will be made to present some of the main themes in the evolution of our concepts on UFO propulsion and related matters.

Early ufologists were immediately impressed by the ability of UFOs to hover, rapidly accelerate, and perform violent acrobatic maneuvers without a visible means of propulsion. They noted that the ones whose landing marks indicated weights in the tens of tons seemed to also have secondary effects involving magnetism, and it appeared logical enough to them that powerful magnetic fields were the source of UFO propulsion. For example, on June 24, 1947 a camper in the Cascade Mountains, Oregon, USA noticed his compass needle spinning wildly about when a UFO overflew his position. These types of cases, coupled with the many in which a UFO witness' wristwatch was stopped due to exposure to a powerful magnetic field, seemed to indicate that magnetic forces were used to move UFOs.

The early UFO contactees were all obliged to strengthen their claims of personal encounters with extraterrestrials by describing the methods of propulsion that they used in their craft. The American contact George Adamski (1891-1965) presented the idea that a flying saucer or "scout ship" could rapidly maneuver about earth's atmosphere by creating a circular electromagnetic field about the ship that would move it by pushing against the earth's weak magnetic field (with an intensity of only about 3/4 of gauss at the mid-northern and mid-southern latitudes.) For travel between planets in our solar system, a mother-ship would employ the same principle except that it would utilize the magnetic fields emanating from our sun and other planets. [1]

Adamski's approach is illustrated and it is obvious that he envisioned the magnetic fields from UFOs interacting with external natural magnetic fields to produce translational motion in a manner similar to the way in which the magnetic fields of a motor's armature windings interact with the magnetic fields of the motor's field windings to produce rotational motion.

The problem with this approach is that it is impossible to produce the necessary magnetic field about the UFO using *any* kind of scheme in which electrical current flows about a complete circuit within the craft. However, even if a magnetic field of the appropriate configuration could be produced, it would have to be enormously powerful to account for the observed acceleration of the UFOs if they are assumed to have their full expected masses. Such intense magnetic fields would not only stall automobiles, but actually twist and deform their steel bodies and chasses as the magnetic fields induced in these structures forced them to align with the UFO's magnetic field. Two UFOs utilizing this form of propulsion would violently smash together if they tried to fly side by side in formation. Also, such a system of propulsion limits the UFO to flying at right angles to the external magnetic field lines the craft utilizes.

Such a propulsion scheme says nothing about the means for producing the craft's powerful magnetic field, nothing about the source of energy which will generate the craft's magnetic field, and nothing about how the craft and crew will be protected from lethal inertial forces while undergoing violent maneuvers. For example, it is known that a right angle turn by a UFO traveling at only 100 miles per hour would kill most human pilots. The magnetic propulsion scheme says nothing about how a UFO crew could be protected from being pulverized while performing a right angle turn at *thousands* of miles per hour. Some early non-contactee ufologists attempted to get around these objections by assuming that all UFOs were automated or carried robot crews. One author, Gerald Heard, provided a unique solution to the problem of inertial forces applied to ufonauts' bodies by rapidly accelerating UFOs.[2] He simply concentrated on the sightings of smaller UFOs (sometimes referred to as "probes") and assumed that they carried intelligent bees from Mars with hard crystal-like bodies that would not be pulverized during violent aerial maneuvering. Perhaps he thought the humming sounds associated with many UFOs were produced by their tiny insect crews!

If UFOs are assumed to have their full, normal masses in flight, then simple calculations show that they would have to have enormously powerful engines to operate as they do. For example, suppose that a witness reported spotting a saucer hovering near ground level that was 32 feet in diameter and which suddenly rose and disappeared into a cloudless sky in only ten seconds. Calculations show that the UFO would have

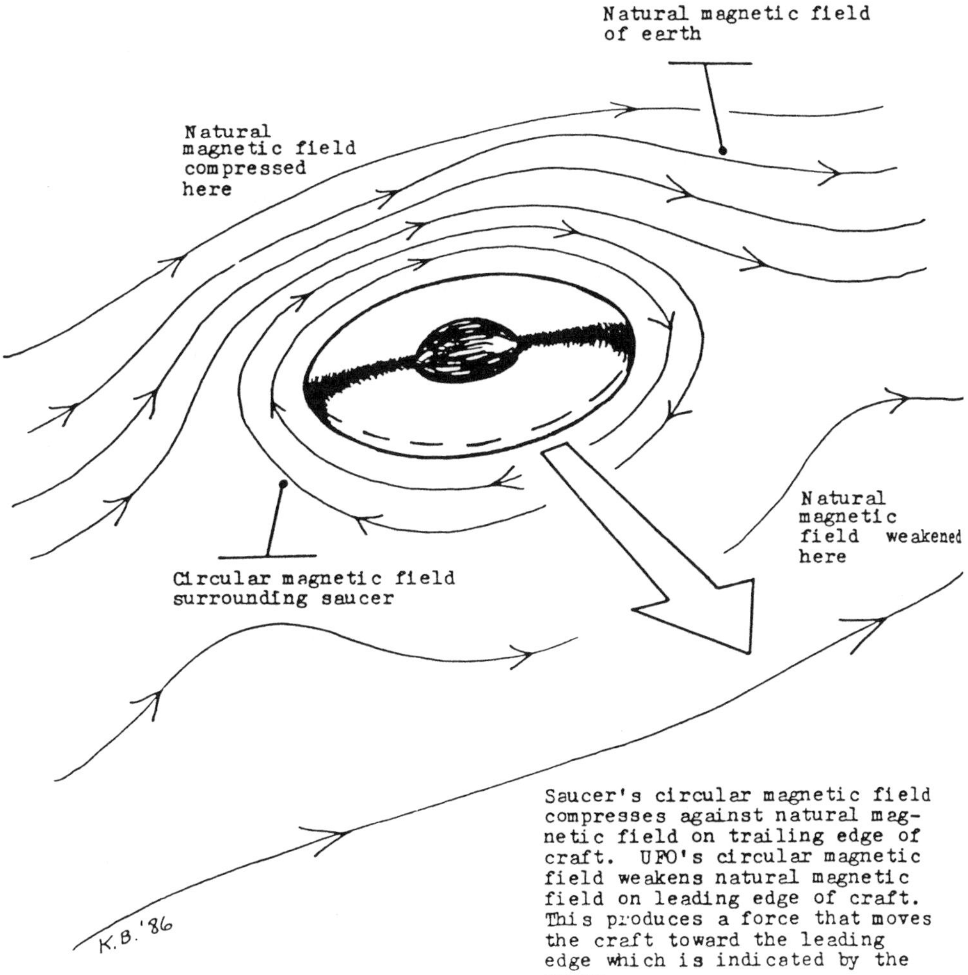

Natural magnetic field
of earth

Natural
magnetic field
compressed
here

Natural
magnetic
field weakened
here

Circular magnetic field
surrounding saucer

K.B. '86

Saucer's circular magnetic field compresses against natural magnetic field on trailing edge of craft. UFO's circular magnetic field weakens natural magnetic field on leading edge of craft. This produces a force that moves the craft toward the leading edge which is indicated by the arrow.

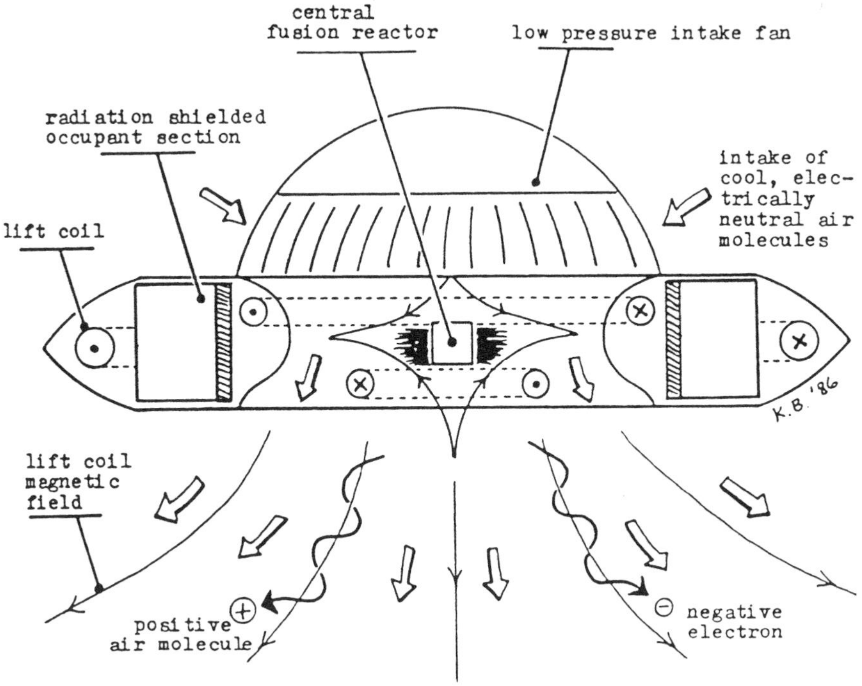

central
fusion reactor

low pressure intake fan

radiation shielded
occupant section

intake of
cool, elec-
trically
neutral air
molecules

lift coil

K.B. '86

lift coil
magnetic
field

positive
air molecule

negative
electron

A powerful exhaust of hot plasma is ejected below the craft in pulses to provide a lifting thrust to rise to an altitude of 20.83 miles in ten seconds to shrink to a point too small for the human eye to resolve. If the craft's acceleration was constant, then its crew would have felt the crushing inertial force produced by an acceleration of 68.38 g's and at the end of its ten-second climb the craft would be moving at 15,000 miles per hour. If the UFO weighed 10 tons (or 20,000 pounds), its engines would have to produce a force of 1.368 *million* pounds. Its engines would have to have had a power output of 27,36 *million* horsepower or 20,000 *mega*watts. Thus its power output would be equivalent to about 2,500 of the 11.000 horsepower Pratt & Whitney jet engines that power the American B-52 Stratofortress bomber or to about twenty 1,000 megawatt nuclear power plants operating at peak output!

When we realize that such power would have to be safely generated within the limited confines of a small vehicle interior and would have to operate without any apparent means so that the UFO's hull can be a single, closed metal surface, we can appreciate the formidable challenge that a working UFO propulsion hypothesis presents to the theorist *if UFOs are considered to be massive in flight.* Most orthodox scientists who have reviewed these kinds of calculations, and assumed that UFOs have their full, normal masses in flight, have tended to become skeptical about the existence of genuine extraterrestrial UFOs in general.

Early attempts to rescue the magnetic propulsion hypothesis and provide the enormous power levels required to move massive craft were made by such researchers as the Canadian engineer Wilbert B. Smith (died 1962). He was the director of Canada's Project Magnet in 1949, which was one of the few open government efforts to seriously study UFOs. Concentrating on the small percentage of UFOs that display rotating hull surfaces, he proposed that their motion would cut the earth's geomagnetic field lines to produce large quantities of electrical power. Thus he likened rotating UFOs to giant electrical generators, whose power output would then be used to heat and expel air or another ejectable mass carried by the craft for thrust. He could not explain how a rapidly moving UFO could protect itself from being heated to incandescence by air friction, and assumed that the craft's rotating motion protected it. To overcome the problems inertial forces present to pilots aboard violently manor UFOs, Smith assumed that all UFOs were automated. Calculations of the amount of rotation of the craft's hull surfaces needed to generate the enormous power levels massive UFOs would need, indicate that a saucer using Smith's propulsion hypothesis would tear itself apart with centrifugal forces.

Toward the end of the 1950s and beginning of the 1960s, UFO propulsionists began to search for other than conventional ways to explain UFO performance and generate the huge power levels that *apparently* massive UFOs would need to achieve anomalous flight. They turned their attentions toward the use of such exotic energy sources as cosmic energy and nuclear power such as fusion.

As explained in a book by Aime Michel,[3] an engineer of the French Air Force, Lieutenant Jules Plantier, had been working on a novel hypothesis for UFO propulsion as early as 1953. Plantier's approach assumed that all space is permeated by a form of high-powered energy that was the source of cosmic rays. The ufonauts had simply developed equipment that could convert this sea of energy into motive force that would power a massive UFO in a planet's atmosphere or in the space between star systems.

In liberating this cosmic energy for motion, a UFO would be surrounded by a *local* electromagnetic field that could be varied and applied at will and which would account for all of the secondary effects associated with UFO propulsion.

Plantier's hypothesis was the first to attempt to explain the color changes associated with changes in UFO motion. Mathematically his hypothesis could indicate that rapid accelerations would envelop a UFO in a red glow, a high-speed turn would produce a green glow, and hovering would produce a white glow.

To explain how UFOs overcome aerodynamic heating

during high-velocity flight, the Plantier hypothesis assumed that the UFO's local EM field would drag some of the air surrounding the craft along with it to act as an insulation cushion.

It is also significant to note that not only did Plantier's approach rationalize several of the secondary effects associated with UFO propulsion, but he also suggested that the universal energy tapped by UFOs would interact and selectively cancel out the earth's or the universe's pull on a massive UFO so as to allow it to move in any direction due to the gravitational pull of the earth or universe still allowed to affect the craft through its local EM field.

Plantier's hypothesis was ultimately unsatisfactory because it did not give the details of how the ufonauts would use cosmic energy to selectively "shield" their craft form the gravitational pulls exerted on it and is thereby incapable of experimental verification or denial. However, it is historically interesting as the first hypothesis that attempted to *mathematically* rationalize the secondary effects noted for UFOs. It also suggested a way to protect ufonauts from inertial forces since the accelerating forces are applied equally to the massive craft and its crew.

Discouraged by the vague nature of such concepts as "universal seas of energy," many propulsionists chose to give their hypothetical saucers more understandable power sources.

One propulsionist, R.H.B. Winder, in a series of articles for *Flying Saucer Review,* offered his design for a discoidal craft that would be able to hover and undergo a vertical acceleration of 30 g's. As shown, the Winder vehicle was slightly over 30 metres (100 feet) in diameter and weighed 1000 tons. It had a raised dome that was really a high pressure fan that would draw in cool, electrically neutral air from ablove the vehicle and force it to flow over a fusion reactor in the center of the hull. Here the air would be ionized by powerful neutron and gamma radiations from a fusion reactor and would also serve to cool the reactor.

Once ionized, the positively and negatively charged air

particles would be momentarily held in cylindrically symmetrical, cusped containment magnetic fields (created by two containment cells with opposite current flows that contain the D-He^3 plasma undergoing fusion) before leaking down into the region below the craft. At this point, a large 3-meter-diameter lift coil would be activated so the magnetic field would slowly pulse from a central field strength of 100 kilogauss (about 200,000 times as strong as the earth's magnetic field) to one of 200 kilogauss. These high-power magnetic fields were achieved by using super conducting niobium/tin electromagnets that could carry hundreds of thousands of amperes at low temperatures.

As the lift coil's magnetic field intensified in strength, it would compress the ionized air trapped in its magnetic field lines and this air was then pushed out below the craft to provide the enormous thrust needed to accelerate such a massive object. Winder calculated that if the lift coil pulsed 1.4 times per second, then each pulse would expel 135 tons of air at a velocity of 680 feet per second. A force of 30,000 tons would then be transmitted back to the craft through the lift coil's magnetic field to accelerated the craft at 30 g's. The craft could also be made to hover if there was one pulse every 7.5 seconds that expelled 1,800 tons of air at a velocity of 40 meters per second.

Winder's design suffered from many problems. Aside from the shielding that would be needed to protect the craft's occupants from the lethal radiations of a centrally-located reactor, there was the problem that the craft would pose for ground crews or observers who would be exposed to 130 km/h blasts of superheated air that would be hot enough to ignite the surface of a tarmacadam road!

Additionally, a craft using this method of propulsion would, when hovering, rise and fall about 30 meters with each pulse of the lift coil's field. Needless to say, this would make a controlled and safe landing of the craft virtually impossible, while the powerful magnetic fields involved could have catastrophic results for several craft flying in close formation.

Most importantly, Winder's propulsion hypothesis said nothing about protecting the crew from the effects of high g acceleration. Because he assumed that UFOs have their full mass in flight, he was forced to use a fusion reactor with a sufficiently high power output (10,200 megawatts for hovering and 12,200 megawatts for full acceleration at 30 g's) to drive his hypothetical 1000-ton ship. Also, his craft was incapable of space-flight unless it could be modified to carry an enormous amount of material to expel for thrust. No means was provided to protect against or eliminate aerodynamic heating of its surfaces in flight, even though other ufologists such as the American Coral Lorenzen were suggesting that UFOs eliminate aerodynamic drag by ionizing the air surrounding their hulls, deflecting it away from the craft with magnetic fields.[5]

In an effort to minimize the power needed to propel a UFO as it overcame the forces of gravity and to protect its crew from inertial forces, the propulsionists of the 1960s turned to a variety of anti-gravity hypotheses that would all eventually prove unsatisfactory.

The English engineer Leonard G. Cramp offered his famous "G-field" hypothesis in two books.[6,7] He claimed that the occupants of a rapidly-accelerating UFO felt no inertial forces acting on their bodies because all the atoms in their bodies as well as in their craft accelerated *at the same rate* in an artificial local gravity field produced around the craft by a set of "G-field projectors" that it carried.

His hypothesis does allow a UFO to overcome gravity and inertial effects while still allowing the craft and crew to have normal masses in flight. He unfortunately could provide no details on how such G-field projectors would operate. Also such a system would probably be unusable for the same reasons that fixing a powerful magnet in from of a small steel car will not produce motion in the car as long as the magnet is physically attached to the car.

Researchers such as the American physicist T. Townsend Brown[8] had developed electro-gravitic devices wherein it was

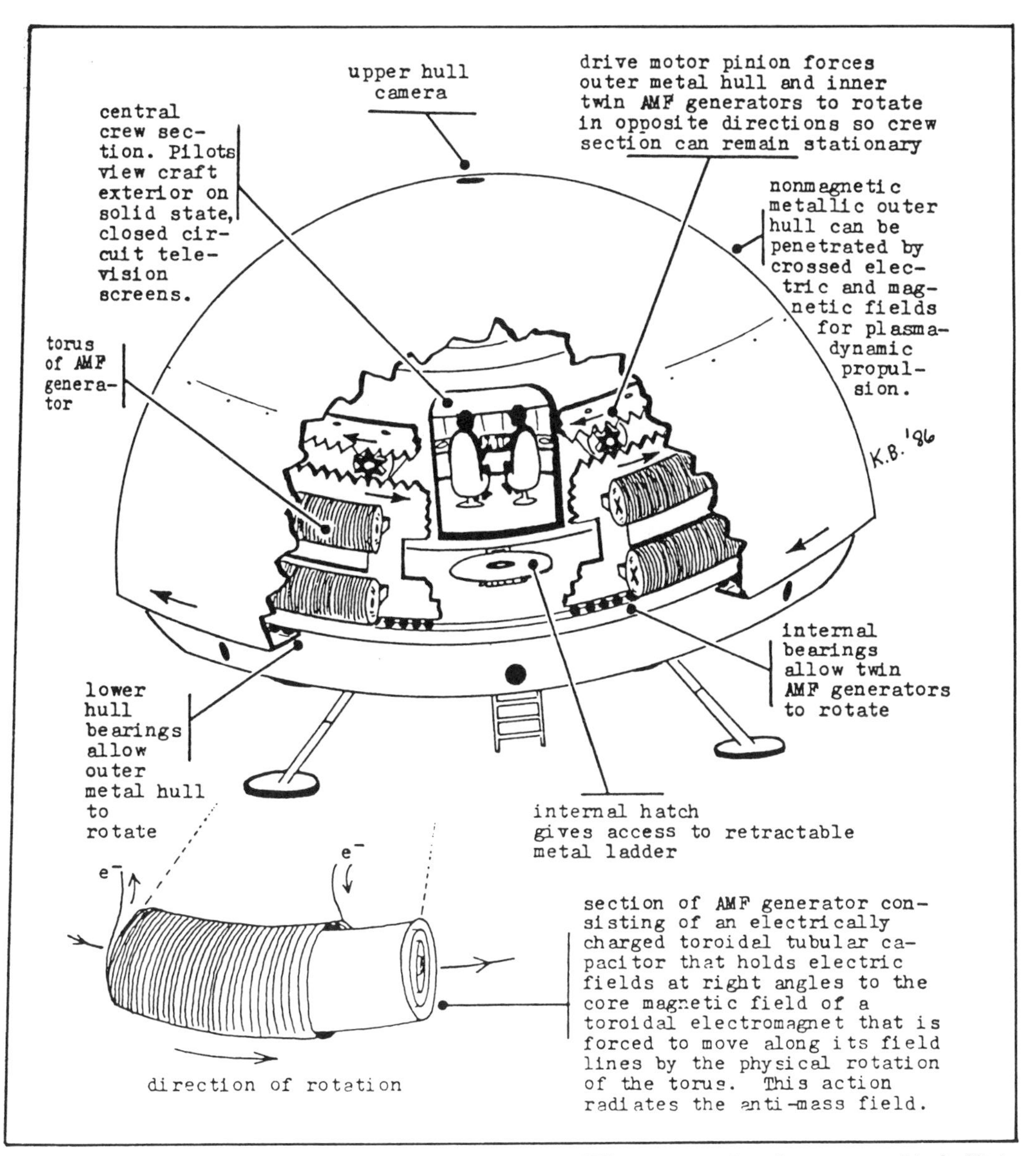
upper hull camera
drive motor pinion forces outer metal hull and inner twin AMF generators to rotate in opposite directions so crew section can remain stationary
central crew section. Pilots view craft exterior on solid state, closed circuit television screens.
nonmagnetic metallic outer hull can be penetrated by crossed electric and magnetic fields for plasmadynamic propulsion.
torus of AMF generator
K.B. '86
internal bearings allow twin AMF generators to rotate
lower hull bearings allow outer metal hull to rotate
internal hatch gives access to retractable metal ladder
e⁻
e⁻
section of AMF generator consisting of an electrically charged toroidal tubular capacitor that holds electric fields at right angles to the core magnetic field of a toroidal electromagnet that is forced to move along its field lines by the physical rotation of the torus. This action radiates the anti-mass field.
direction of rotation

Rotating hemispherical UFO equipped with twin AMF generators (sections removed to facilitate view of interior). This craft is designed for atmospheric operation only and is carried by a mothership.

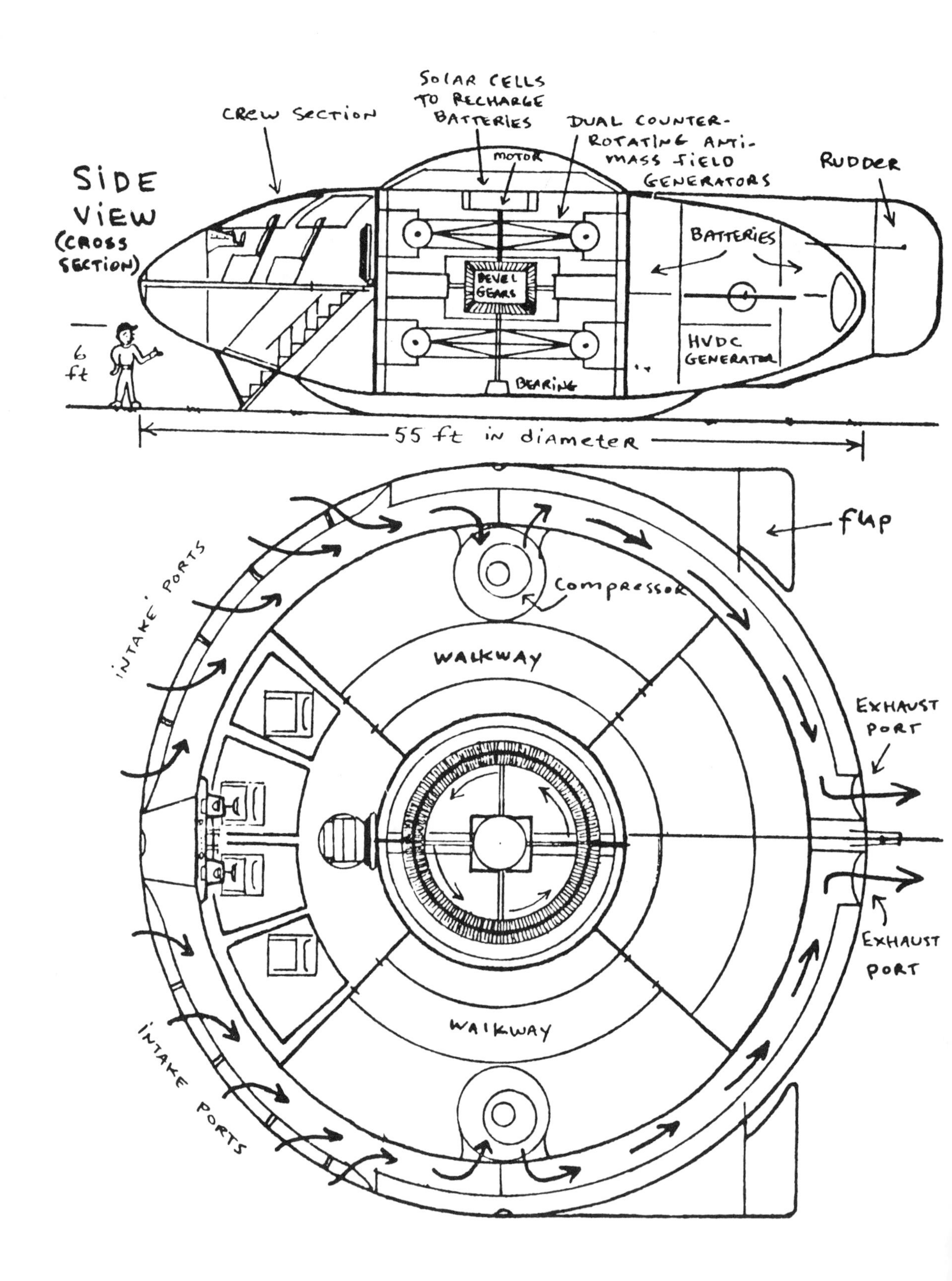
SIDE VIEW (CROSS SECTION)
CREW SECTION
SOLAR CELLS TO RECHARGE BATTERIES
MOTOR
DUAL COUNTER-ROTATING ANTI-MASS FIELD GENERATORS
RUDDER
BATTERIES
BEVEL GEARS
HVDC GENERATOR
BEARING
6 ft
55 ft in diameter
flap
compressor
intake ports
WALKWAY
EXHAUST PORT
WALKWAY
EXHAUST PORT
intake ports

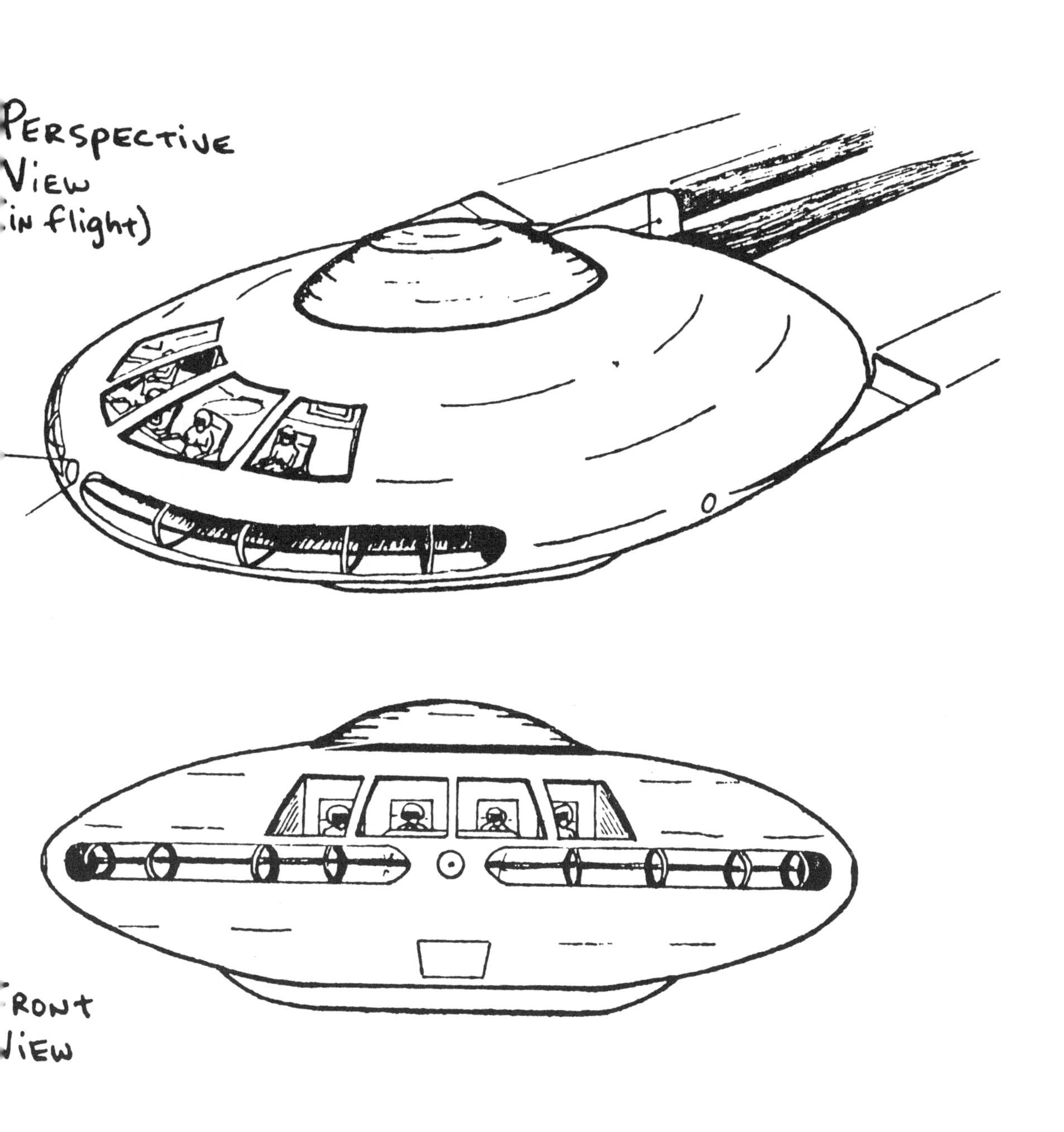
PERSPECTIVE
VIEW
(IN flight)
FRONT
VIEW

claimed that certain high-capacitance dielectrics would exhibit a propulsive force toward their positively-charged plates when subjected to a high voltage of 50 to 300 kilovolts. Small suspended discs utilizing this effect were capable of motion, but orthodox physicists tended to dismiss the phenomenon as due to "electric wind" which was just a flow of ionized air around the highly-charged discs. Although they glowed like real UFOs and were capable of motion, no experiments have ever indicated that any electron device had experienced a loss of mass in operation.

During the 1970s most of the UFO propulsionists had grown disenchanted with the anti-gravity approach and began to finally abandon the extraterrestrial hypothesis because of the failure of its adherents to develop a working hypothesis that would allow the craft's performance and secondary effects to be duplicated by earthly versions. It is at this point in the history of the subject that many embraced the extradimensional hypothesis which avoided the problems of vehicle propulsion by assuming that the craft were not really spacecraft as we understood the term, but rather devices for traveling through space/time or from a higher or lower set of dimensions into our four-dimensional universe.

Researchers such as the American Alan C. Holt[9] proposed a unified-field-theory model that attempted to describe all matter and energy in the universe as the manifestation of hyperspace energy currents. According to his hypothesis, a UFO could change its location by converting electromagnetic energy into gravitational energy. Using laser pulses, magnetic fields, and electron beams, his "resonance spacecraft" would generate oscillating energy patterns in a metal torus that would then interact with the postulated hyperspace energy patterns to negate the effects of local gravity fields. He envisioned a craft which, when only slightly out of resonance with its local hyperspace energy pattern, would perform the violent anomalistic motions sometimes noted for UFOs. When greatly out of resonance with its local hyperspace environment, his craft would then instantly travel through space/time to its resonant position in the universe, which could be very far from its starting point.

Another American researcher, Daniel Eden, has proposed a novel explanation for the "multiple image" UFO photographs that apparently show the UFO materializing in several different locations during the brief time that the camera's shutter is open.[10] He proposed that these photographs demonstrate that the craft is actually rapidly oscillating or "coining" from a dimensional space/time continuum higher than ours to one lower than ours. Thus, he solves the problem of overcoming gravity and inertia in our space/time continuum by having the UFO's pilots constantly adjusting the location in our universe that the craft passes through.

There are many problems with the extradimensional approach to UFO propulsion and dynamics. They all rely on the assumed existence of such things as parallel universes and an array of vague and as yet undiscovered effects that would allow access to these assumed realms. They say nothing definite about the many secondary effects such as automobile engine stalling, the extinguishing of incandescent electric lamps, interference with radio and television reception and power distribution, and the levitation of automobiles and drivers near large hovering UFOs. Neither does this approach explain the alarming physiological effects noted, such as the skin burns and paralysis sometimes received by human witnesses in close proximity to UFOs.

During the 1970s and 1980s, only a handful of ufologists had remained faithful to the extraterrestrial hypothesis and sought materialistic solutions for the phenomenon that would be capable of experimental verification.

The American ufologist James M. McCampbell proposed the idea that UFOs were material craft whose builders had developed a technology that allows them to greatly reduce the mass of objects.[11] He has attempted to rationalize the various colored glows surrounding the craft in terms of the irradiation of the air surrounding the vehicle by powerful microwave radiations it was emitting.

Although he has not provided a mechanism to explain how the craft would lower its mass so as to reduce its gravitational and inertial properties, his research into UFO microwave emissions has successfully explained the so-called "hot ring" effect. This effect is sometimes noted by close encounter witnesses when they feel their finger rings getting hot. Apparently the rings can act as miniature loop antennas that can turn received microwave radiation into electrical currents that will heat up a metallic ring.

This writer has also remained faithful to the ETs and believes that we will only be able to eventually duplicate the propulsion systems utilized by the UFOs and be able to reproduce the secondary effects associated with this technology, if we assume UFOs are physical craft that engage in travel between star systems by moving through *our* universe in a manner that uses a technology only a *little* more advanced than what we already possess!

During the last decade, this writer has managed to systematically develop a new hypothesis for UFO propulsion and secondary effects which successfully meets all the criteria for a usable working hypothesis. This new hypothesis is referred to as the "anti-mass field theory" or "AMF theory" although, technically, it is still closer to being a hypothesis than a full-fledged theory.

It begins with a new model to describe how objects produce the gravitational and inertial properties which we measure and use to define the "mass" of the object. The new model assumes that all of the sub-atomic particles in the atoms that comprise an object emit or radiate a kind of non-electromagnetic radiation that expands away from the object at the speed of light. It is the refraction or bending of this radiation as it interacts with the radiation emitted form other objects which produces the force we call "gravity" that tends to pull all objects together. When an object moves, its motion forces the invisible radiation it emits to bend and this bending creates the force we call "inertia" which tends to retard or oppose the acceleration and deceleration of objects. Contemporary physicists may refer to this unseen radiation emitted form the sub-atomic particles of an object as the object's "gravity field." This writer, however, prefers to call it the object's "mass field."

From the above it should be obvious that if we could somehow partially reduce or partially neutralize the mass field emitted from an object, then that object would lose some of its gravitational and inertial properties and would *effectively* have a lower mass. If all the mass field emitted from an object could be extinguished, then that object would become totally massless.

The beings who designed and built the extraterrestrial craft that visit our world have learned how to neutralize the normal mass fields emitted from the atoms of their craft and crews' bodies. According to AMF theory, this is accomplished by equipping their craft with devices known as "anti-mass field generators." These devices emit or radiate a field whose "polarity" is opposed to that of the craft and crew's normal mass field and which cancels out or negates this mass field. With little or no mass, a UFO and its crew can buoyantly float in a planet's atmosphere without the enormous expenditures of energy that massive craft require. They can also travel through interstellar and intergalactic space at velocities exceeding that of light while massless without violating Einsteinian relativity. [12, 13]

AMF theory goes on to provide a *detailed* mechanism for the means by which an anti-mass field is generated and a general description of the equipment needed to produce this new field effect. Basically, as shown, the AMF generator built into a UFO emits its anti-mass field when a magnetic field is forced to move along its field lines at right angles to an electric field. AMF generators are usually toroidal or donut-shaped and consist of a toroidal electromagnet that produces a powerful magnetic field at the core of the torus. The core also contains a toroidal tubular capacitor that consists of charged toroidal plates that hold electric fields at right angles to the core's magnetic field.

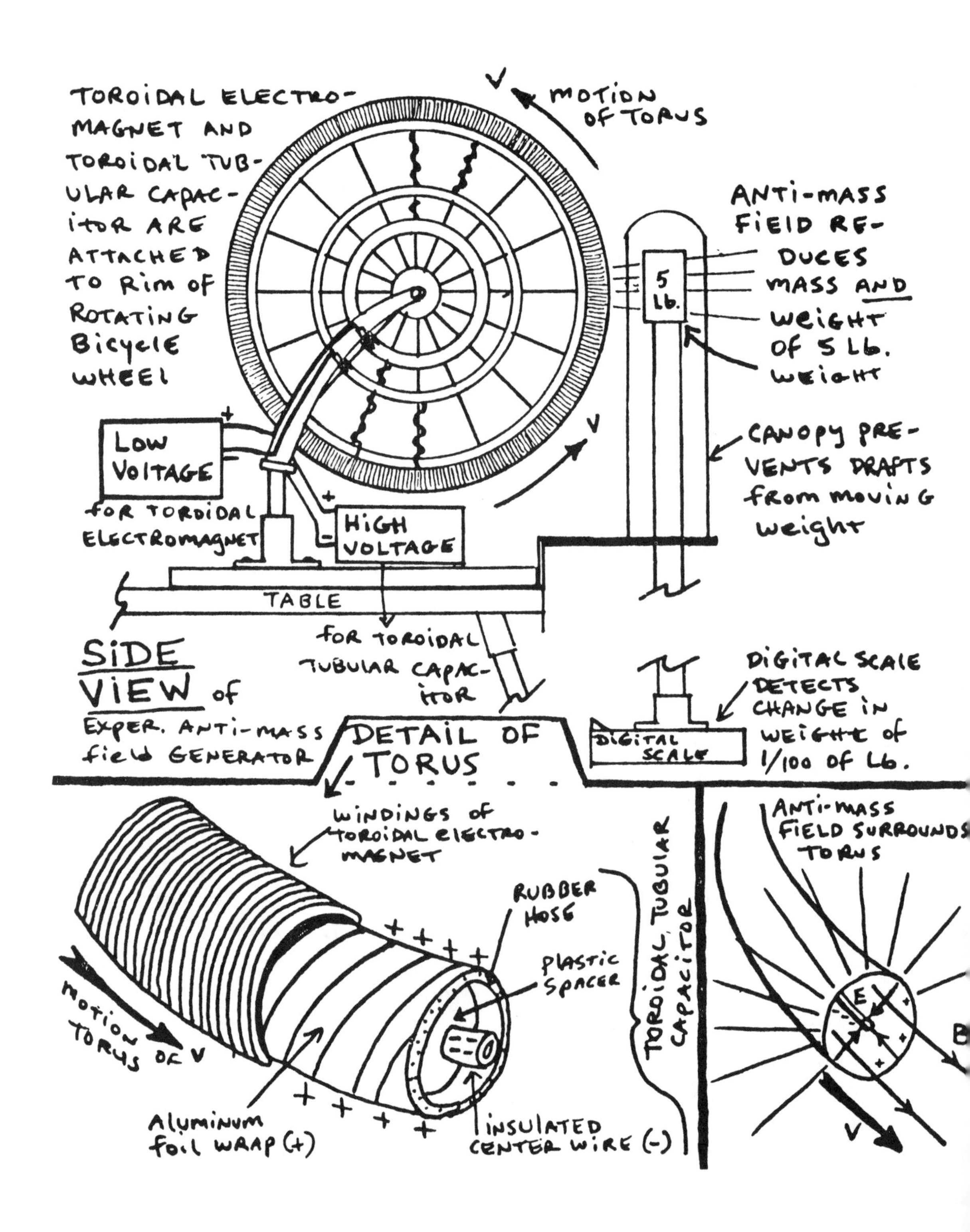
TOROIDAL ELECTRO-MAGNET AND TOROIDAL TUBULAR CAPACITOR ARE ATTACHED TO RIM OF ROTATING BICYCLE WHEEL
V
MOTION OF TORUS
ANTI-MASS FIELD REDUCES MASS AND WEIGHT OF 5 LB. WEIGHT
5 LB.
V
CANOPY PREVENTS DRAFTS FROM MOVING WEIGHT
LOW VOLTAGE
FOR TORDIDAL ELECTROMAGNET
HIGH VOLTAGE
TABLE
FOR TOROIDAL TUBULAR CAPACITOR
SIDE VIEW of EXPER. ANTI-MASS FIELD GENERATOR
DETAIL OF TORUS
DIGITAL SCALE
DIGITAL SCALE DETECTS CHANGE IN WEIGHT OF 1/100 OF LB.
WINDINGS OF TOROIDAL ELECTRO-MAGNET
RUBBER HOSE
PLASTIC SPACER
MOTION OF TORUS V
ALUMINUM FOIL WRAP (+)
INSULATED CENTER WIRE (-)
TOROIDAL TUBULAR CAPACITOR
ANTI-MASS FIELD SURROUNDS TORUS
E
B
V

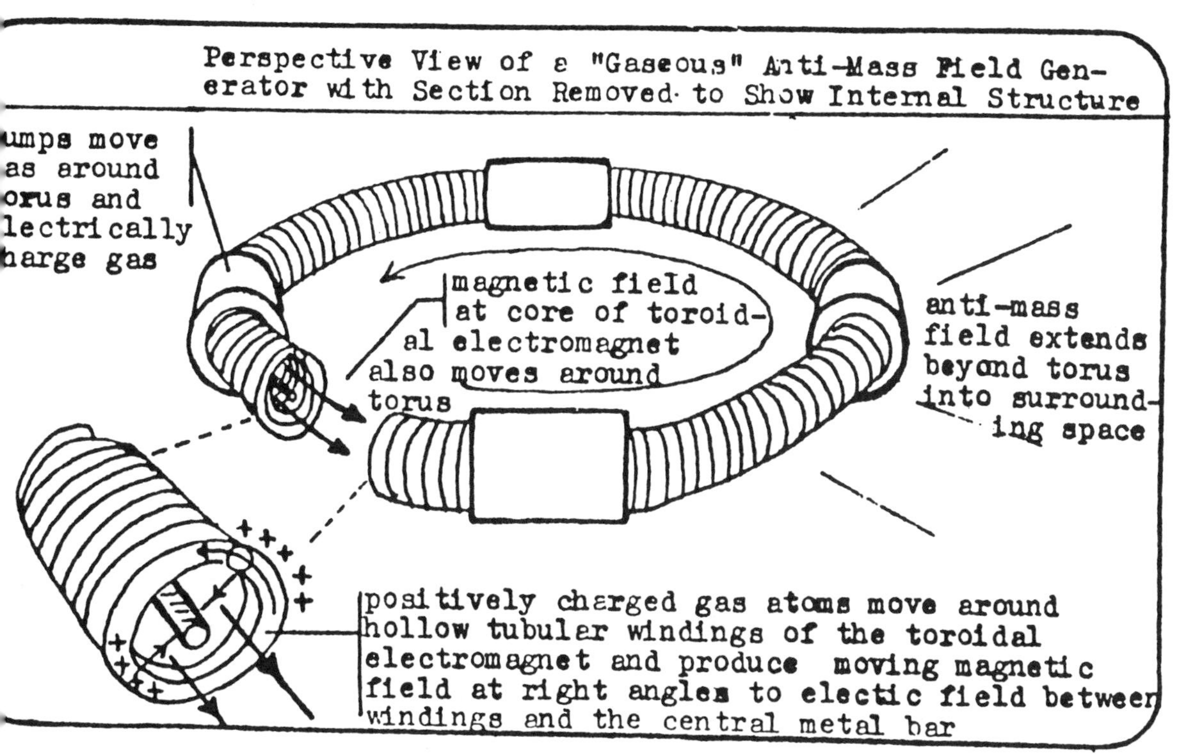
umps move
as around
orus and
lectrically
harge gas
magnetic field
at core of toroid-
al electromagnet
also moves around
torus
anti-mass
field extends
beyond torus
into surround-
ing space
positively charged gas atoms move around
hollow tubular windings of the toroidal
electromagnet and produce moving magnetic
field at right angles to electic field between
windings and the central metal bar
Perspective View of a "Gaseous" Anti-Mass Field Generator with Section Removed to Show Internal Structure

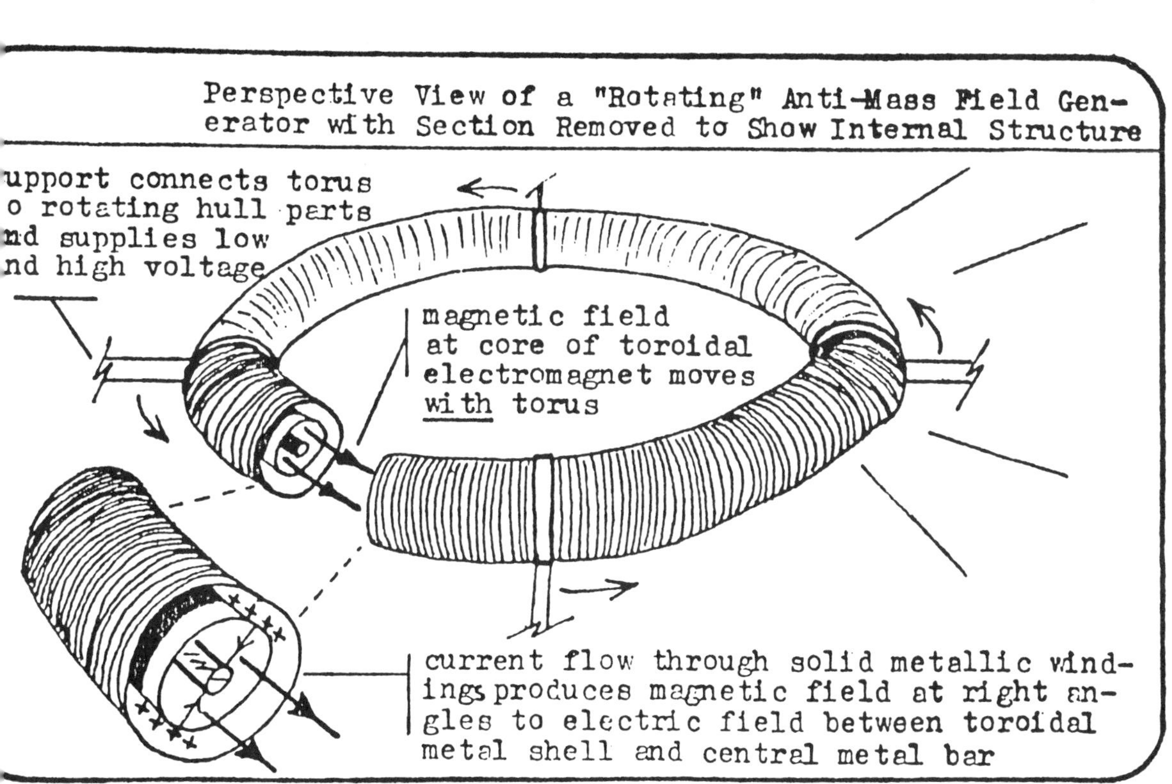
upport connects torus
o rotating hull parts
nd supplies low
nd high voltage
magnetic field
at core of toroidal
electromagnet moves
with torus
current flow through solid metallic windings produces magnetic field at right angles to electric field between toroidal metal shell and central metal bar
Perspective View of a "Rotating" Anti-Mass Field Generator with Section Removed to Show Internal Structure

The magnetic and electric fields inside the toroidal AMF generator are not sufficient by themselves to generate an anti-mass field. It is still necessary to move the core's magnetic field along its field lines, and this is usually achieved by *physically* rotating the entire device. This requirement is obvious from the many UFOs that display rotating hull surfaces. In these cases, the craft's AMF generators are attached mechanically to its rotating hull surfaces and rotate counter to them inside the hull. When such a craft has landed, it will slow and stop the rotation of its AMF generators and regain its full mass. Since it now has gravitational and inertial mass again, it may weigh tens of tons and press its landing gear into the soil of the landing site. Prior to lift-off, the craft will increase the rotation rate of its hull surfaces, and their internally moving AMF generators will emit its anti-mass field. The craft will then become almost massless and can rapidly leave the site when propulsive forces are applied to its hull. No matter how violent the take-off, the crew will feel little if any inertial forces applied to their nearly-massless bodies.

When operating in earth's atmosphere, some UFOs use simple rocket thrusters to apply motive force to their hulls. Because the craft is buoyant due to its AMF, its thrust can be used solely to provide horizontal motion. Such craft, however, are limited to velocities of only a few thousand km/h if excessive hull heating is to be avoided.

A more sophisticated means of thrust used by most UFOs is referred to as the "plasmadynamic mode of propulsion" in AMF theory. A plasma dynamically-propelled UFO uses its anti-mass field and various magnetic fields emanating from a *non-magnetic* metallic hull to convert the layer of air immediately surrounding its hull into a rich plasma or mixture of positively-charged atmospheric ions and negatively-charged electrons. This rich plasma is then exposed to crossed or perpendicular electric and magnetic fields that are projected into the plasma from propulsion equipment located inside the UFO's hull. These crossed electric and magnetic fields then apply what are known as "Lorentz forces" to the charged plasma particles and force them to flow *around* the craft's hull at enormous velocity. Since the plasma is forced to flow around the hull without touching it, aerodynamic drag is virtually eliminated and a massless UFO can achieve extremely rapid accelerations in our atmosphere without producing turbulence such as the "sonic boom."

AMF theory can further be used to explain how the boundary layer of plasma surrounding a moving UFO can be made to emit a visible glow. As the plasma's charged particles move through the crossed electric and magnetic fields near the craft's hull, the charged particles will move with an odd cycloidal motion. This motion forces them to emit a form of electromagnetic radiation known as "cyclotron radiation." Calculations show that if the magnetic fields near the hull have a strength of about 1000 gauss and the masses of the electrons in the plasma are reduced to about one ten-thousandth of their normal values by the UFO's anti-mass field, then these electrons can emit *visible* cyclotron radiation in the infrared and ultraviolet regions of the spectrum. It is these radiations which cause the heating, skin burning, and eye injury secondary effects associated with UFO propulsion systems.

The AMF theory has, so far, successfully explained virtually all of the UFO-related secondary effects that it has been applied to. For example, it provides a mechanism to account for the way UFOs cause automotive failures and power blackouts. In these cases distant magnetic fields can interact with the UFO's anti-mass field reaching them to ionize air near the magnetic fields. This ionized air is electrically conductive and can short-circuit high-voltage circuits such as automobile ignition systems and high-tension power lines near their support towers.

AMF theory is still in a process of development and refinement, but it is *already* sufficiently detailed enough to allow us to begin the research needed to duplicate UFO propulsion.

References

1. Adamski, George. *Inside the Spaceships.* Abelard-Schuman, London, 1955.
2. Heard, Gerald. *Is Another World Watching? The Riddle of the Flying Saucers.* Bantam Books, New York, 1953.
3. Michel, Aime. *The Truth About Flying Saucers.* Criterion Books, New York, 1956.
4. Winder, R.H. B. "Design for a Flying Saucer," *Flying Saucer Review.* 1966, Vol. 12, No. 6, p. 21; 1967, Vol. 13, No. 1, p. 13; 1967, Vol. 13, No. 2, p. 20; Vol 13, No. 3, p. 9.
5. Lorenzen, Coral E. *The Great Flying Saucer Hoax,* William Hendricks Press, New York, 1962, p. 139.
6. Cramp, Leonard G. *Space, Gravity and the Flying Saucer.* The British Book Center, New York, 1955.
7. Cramp, Leonard G. *Piece for a Jigsaw.* Sommerton Publishing Company, Cowes, Isle of Wight, England, 1966.
8. Moore, William. "The Wizard of Electrogravity Revisited," UFO Report, Winter 1981, Vol. 9, No. 4, p. 42, New York.
9. Holt, Alan C. "Starcraft: The Anti-Gravity Fleet," *Science Digest,* May 1982, p. 78, New York.
10. Eden, Daniel. "Effective Mass and the UFO," *Pursuit,* Fourth Quarter 1984, Vol. 17, No. 4 (whole No. 68), Little Silver, New Jersey.
11. McCampbell, James M. *Ufology: New Insights from Science and Common Sense.* Celestial Arts, Millbrae, California, 1976.
12. Behrendt, Kenneth. "Introduction to Anti-Mass Field Physics," *AURA,* Aug. 1985, Vol. 1, No. 3, p. 1. Elizabeth, New Jersey.
13. Behrendt, Kenneth. "Beyond Relativity Theory and the Light Barrier," *AURA,* Feb. 1986, Vol. 2, No. 5, p. 1, Elizabeth, New Jersey.

Keneth W. Behrendt has a BA and MS in chemistry from Rutgers University. He has held various technical positions as quality controller in the electronics field, phamaceutical chemist, and engineering consultant for a metal refinery. He has published several articles, including *UFO Propulsion Systems, Origins and purposes* (1978) and in 1985 he commenced publication of *AURA* (Annals of Ufological Research Advances) which treats the anti-mass field theroy in depth. His first book length work *The Physics of the Paranormal,* 1987, applies anti-mass field theory to various paranormal and parapsychological phenomena.
Correspondence to Mr. Behrendt should be sent to: 274 Second St. Elizabeth, New Jersey, 07206 USA.

THOSE CRAZY ANTI-GRAVITY 20s & 30s

Illustration by Mick Brownfield

Nikola Tesla's fantastic Warcliff tower designed to transmit energy through the earth so that any electrical device could receive power ***anywhere***. Note the Anti-Gravity airships hovering around the tower, drawing power from it.

Ere many generations pass our machinery will be driven
by power obtainable at any point in the universe...
it is a mere question of time when men will succeed in
attaching their machinery to the very wheelwork of natured.
— Nikola Tesla, 1891

No other era was more entranced by the vision of anti-gravity and the tremendous potential that it held, than that hopeful time of the 1920's and 1930's. Science fiction had become a popular media from with everything from pulp-fiction magazines to super-science cliff hanger Saturday matinees.

Free energy, magic motors and wireless transmission of power were popular subjects. Nikola Tesla and Albert Einstein had caught the publics imagination, and for a time it seemed that anything was possible. A science fiction reality was about to come into being and change the world for the better. An age of giant airships, powered by Anti-Gravity and free energy devices was only steps away.

Unfortunately, the specter of World War II loomed on the horizon and scientific developments became a matter of great secrecy and military potential. When it was over, Nikola Tesla was dead, dying poor and discredited in a fleabag hotel room in New York. Later, the scientific community would go to drastic steps to try and erase his name from the school books so that only electrical engineers would be aware of him, and even then their study and knowledge of his works would be severely limited. While every American school kid would learn that Otis invented the elevator and Marconi the radio (even though Tesla won a patent dispute with Marconi) Nikola Tesla would fade away into the dim mists of science. His revolutionary inventions would remain science fiction, and that great era dreamed of in the 20's & 30's would once again be science fiction. Today, the dream of anti-gravity airships, free energy motors and wireless transmission of power seem farther away than ever. Or at least that is what the scientific establishment would have us believe.

According to articles written Tesla himself, his wireless transmission and reception is based on the phenomenon of terrestrial resonance, which he discovered in 1899 and his "magnifying transmitter." He considered the whole earth as huge wire or conductor, and having determined its constants in electrical units, he designed the proper wireless transmitter needed to set the globe into powerful electrical vibrations, so that at any point of its surface electricity could be drawn.

Said a 1929 article on Tesla's wireless transmission of power, "If we desire to operate lights or motors according to Tesla's power transmission theory, all we have to do is connect an electrical capacity such as an antenna or other suitable system of conductors through the apparatus. This capacity then absorbs its proper quota of oscillating electrical energy of the transmitter. We have today only radio signals or radio phone speech, but it Tesla's theory is right, and many engineers think is is, eventually we shall have no more power transmission lines running from central stations for hundreds of miles to supply us current for our electric lights, stove, heaters and fans. We shall instead have an electrical capacity in the form of a ball or cylinder perhaps, placed in our attic or possibly in the ceiling of the house and when connected through a Tesla transformer to the earth, we will pick up the desired electrical energy to operate our household devices. It will be a simple matter to connect a meter with this arrangement, so that the energy can be measured and paid for in the regular way to the central station owners."

Illustrated here is Tesla's *World-Wide Transmission of Electrical Signals*. These are drawings from a 1929 article showing the "Theory, Analogy, and Realization. Tesla's experiments with 100 foot discharges at potentials of millions of volts have demonstrated that the recoverable ground waves of Tesla fly through the earth. Radio engineers are gradually beginning to see the light and that the laws of propagation laid down by Tesla over a quarter of a century age form real and true basis of all wireless transmission today."

Back in the twenties anti-gravity airships were all the rage, and most scientists believe the technology to be right around the corner. Nikola Telsa certainly believed so. Buckminster

TRUE REPORTS OF THE STRANGE & UNKNOWN

FATE

January
1990

USA $1.95
CAN $2.75
U.K £ 2

a Llewellyn Publication

TESLA'S DEATH RAY

THE MINISTER OF FAIRYLAND

THE FLYING HORSEMAN

SEEING YOUR FUTURE

The New Age & Beyond

Cover of the November 1929 issue of AIR WONDER STORIES published by Hugo Gernsba
showing a levitated island spaceport city with dozens of anti-gravity airships hovering around it.

According to articles written Tesla himself, his wireless transmission and reception is based on the phenomenon of terrestrial resonance, which he discovered in 1899 and his "magnifying transmitter." He considered the whole earth as huge wire or conductor, and having determined its constants in electrical units, he designed the proper wireless transmitter needed to set the globe into powerful electrical vibrations, so that at any point of its surface electricity could be drawn.

Said a 1929 article on Tesla's wireless transmission of power, "If we desire to operate lights or motors according to Tesla's power transmission theory, all we have to do is connect an electrical capacity such as an antenna or other suitable system of conductors through the apparatus. This capacity then absorbs its proper quota of oscillating electrical energy of the transmitter. We have today only radio signals or radio phone speech, but it Tesla's theory is right, and many engineers think is is, eventually we shall have no more power transmission lines running from central stations for hundreds of miles to supply us current for our electric lights, stove, heaters and fans. We shall instead have an electrical capacity in the form of a ball or cylinder perhaps, placed in our attic or possibly in the ceiling of the house and when connected through a Tesla transformer to the earth, we will pick up the desired electrical energy to operate our household devices. It will be a simple matter to connect a meter with this arrangement, so that the energy can be measured and paid for in the regular way to the central station owners."

Illustrated here is Tesla's *World-Wide Transmission of Electrical Signals*. These are drawings from a 1929 article showing the "Theory, Analogy, and Realization. Tesla's experiments with 100 foot discharges at potentials of millions of volts have demonstrated that the recoverable ground waves of Tesla fly through the earth. Radio engineers are gradually beginning to see the light and that the laws of propagation laid down by Tesla over a quarter of a century age form real and true basis of all wireless transmission today."

Back in the twenties anti-gravity airships were all the rage, and most scientists believe the technology to be right around the corner. Nikola Telsa certainly believed so. Buckminster

Fuller wanted to use anti-gravity airships to unite the world, plant his pre-fabricated 10-story towers, and help bring man into the 21st century with appropriate technology that was enviromentally sound.

Recently, experiements by Tesla with his Warcliff Tower in Long Island were said by Oliver Nichelson in his article on "The Death Ray of Nikola Tesla" (January, 1990 issue of Fate Magazine) to be the cause of the 1908 explosion over Tunguska, Siberia. Nichelson theorized that Tesla wanted to demonstrate for the arctic explorer Peary on Ellesmere Island his ray, but overshot the bleak arctic area Peary was at and instead devastated a remote region of Siberia.

Wave theory was quite in vogue at the time, and radio wave technology held many promises. According to some researchers, Nikola Tesla student Guglielmo Marconi (1874-1937), widely credited with the invention of the radio, though Tesla legally won a patent dispute with him, demonstrated a similar death ray to the Italian dictator Mussolini just prior to his death.

Later, it was alleged by several Italian writers that some of Marconi's followers had built a secret base in the South American jungles (allegedly in a volcanic crater in southern Venezuela) from where they built anti-gravity space ships and conducted secret experiments. One book, written in Italian (and never translated into English, as far as is known) even alleged that the group had taken a trip to Mars in one of the Marconi designed spaceships operating out of Venezuela!

Anti-Gravity was the rage back in the 20s in the 30s, yet the world just wasn't ready for it back then. Maybe the time has finally come...

CHIEF EXPLORER
BUCK ROGERS SOLAR SCOUTS
TO THE FINDER OF THIS BOOK
This Book is the Personal Property of
Name
Address
It contains important secret information of Buck Rogers Solar Scouts. So please return it unopened to me. I will gladly tell you how you can become a member of this club of boys and girls who want to grow up healthy and strong on Cream of Wheat . . . . and have fun doing it.
BUCK ROGERS SOLAR SCOUTS
SECRET CLUB OF THE RADIO FRIENDS OF BUCK AND WILMA

In 1956, Jim Powers of Ford Motor Company drew a "Moonport" as part of the series, "Life in the Year 2000." Most of the picture is still fantasy, but note the surprising resemblance between this spacecraft and today's Concord aircraft. —

Buck Rogers flew around the solar system and galaxy in his own anti-gravity airship that buzzed and flew erratically, much as the Apollo Lunar Module did!

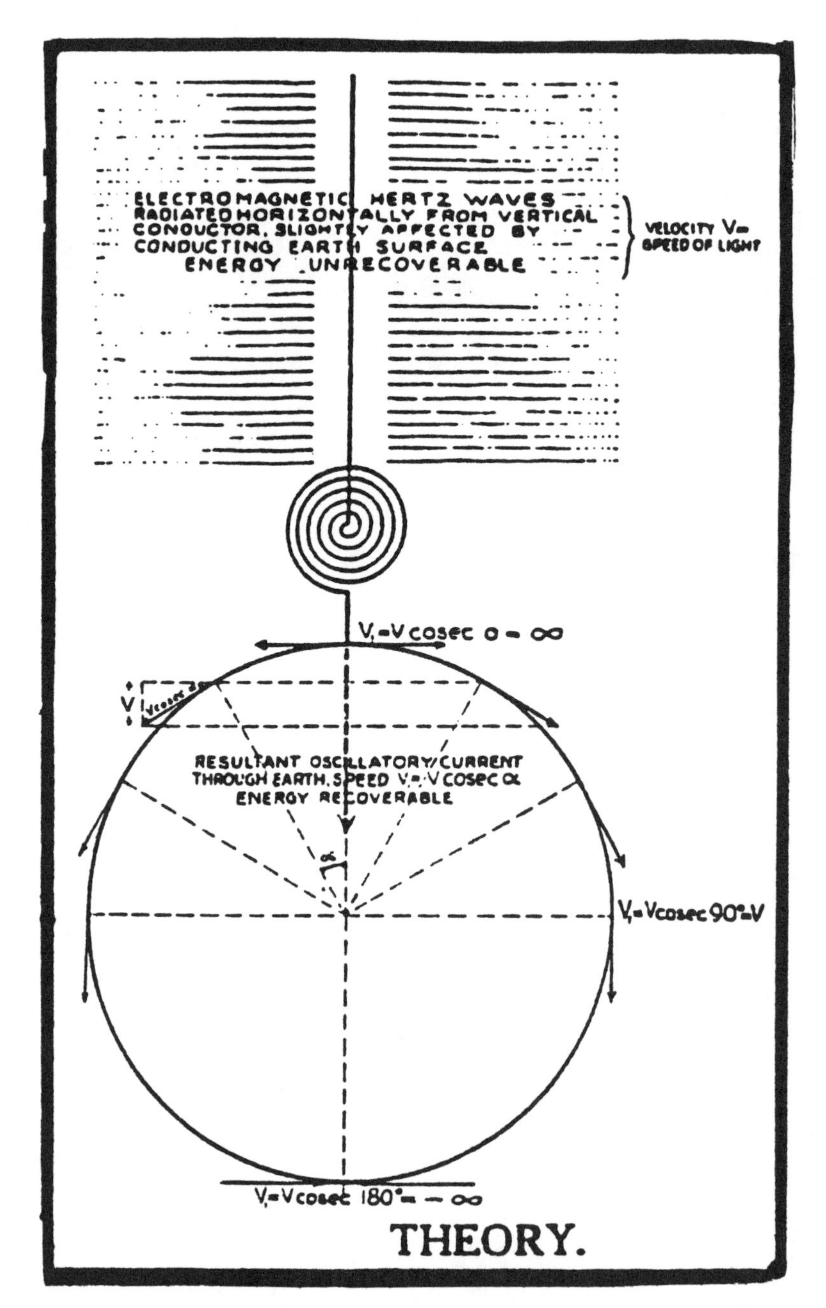

Illustrated here is Tesla's *World-Wide Transmission of Electrical Signals.* These are drawings from a 1929 article showing the "Theory, Analogy, and Realization. Tesla's experiments with 100 foot discharges at potentials of millions of volts have demonstrated that the recoverable ground waves of Tesla fly through the earth. Radio engineers are gradually beginning to see the light and that the laws of propagation laid down by Tesla over a quarter of a century age form real and true basis of all wireless transmission today."

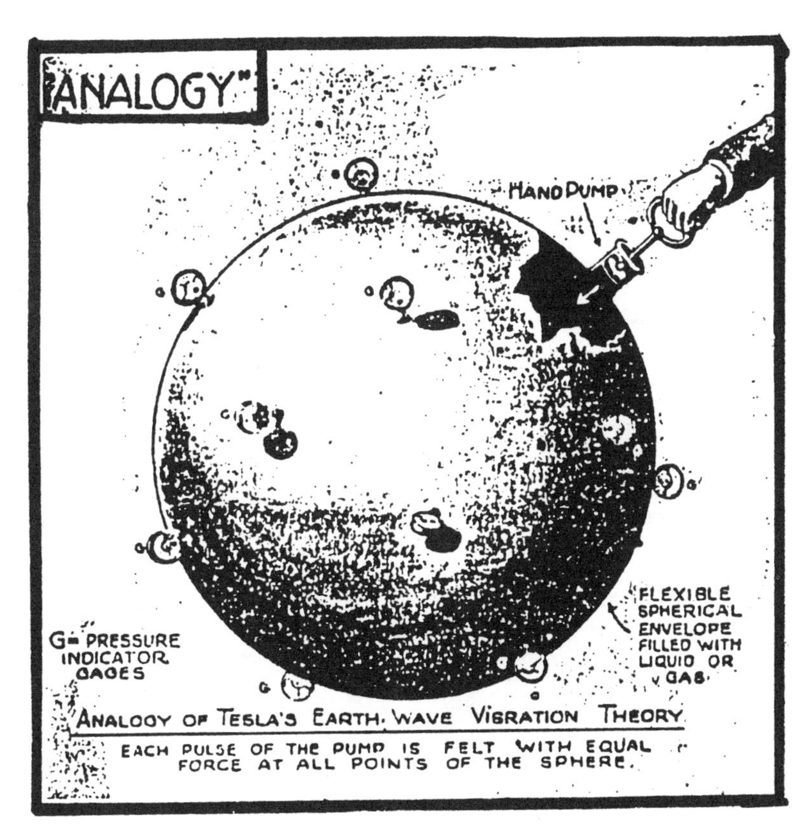
"ANALOGY"
HAND PUMP
G= PRESSURE INDICATOR GAGES
FLEXIBLE SPHERICAL ENVELOPE FILLED WITH LIQUID OR GAS
ANALOGY OF TESLA'S EARTH WAVE VIBRATION THEORY
EACH PULSE OF THE PUMP IS FELT WITH EQUAL FORCE AT ALL POINTS OF THE SPHERE.

"REALIZATION"
THE WIRELESS LIGHT: PLACE A WIRE IN THE GROUND: THAT IS ALL
TESLA'S WIRELESS POWER FOR PROPELLING SHIPS AND AEROPLANES.
TESLA'S WIRELESS TRANSMISSION THEORY.
THE OSCILLATING ENERGY SURGES THRU THE EARTH TO EVERY POINT ON THE GLOBE. THUS ELECTRIC LIGHT, HEAT AND POWER CAN BE DRAWN AT ANY POINT OF THE EARTH FROM A UNIVERSAL CENTRAL STATION.

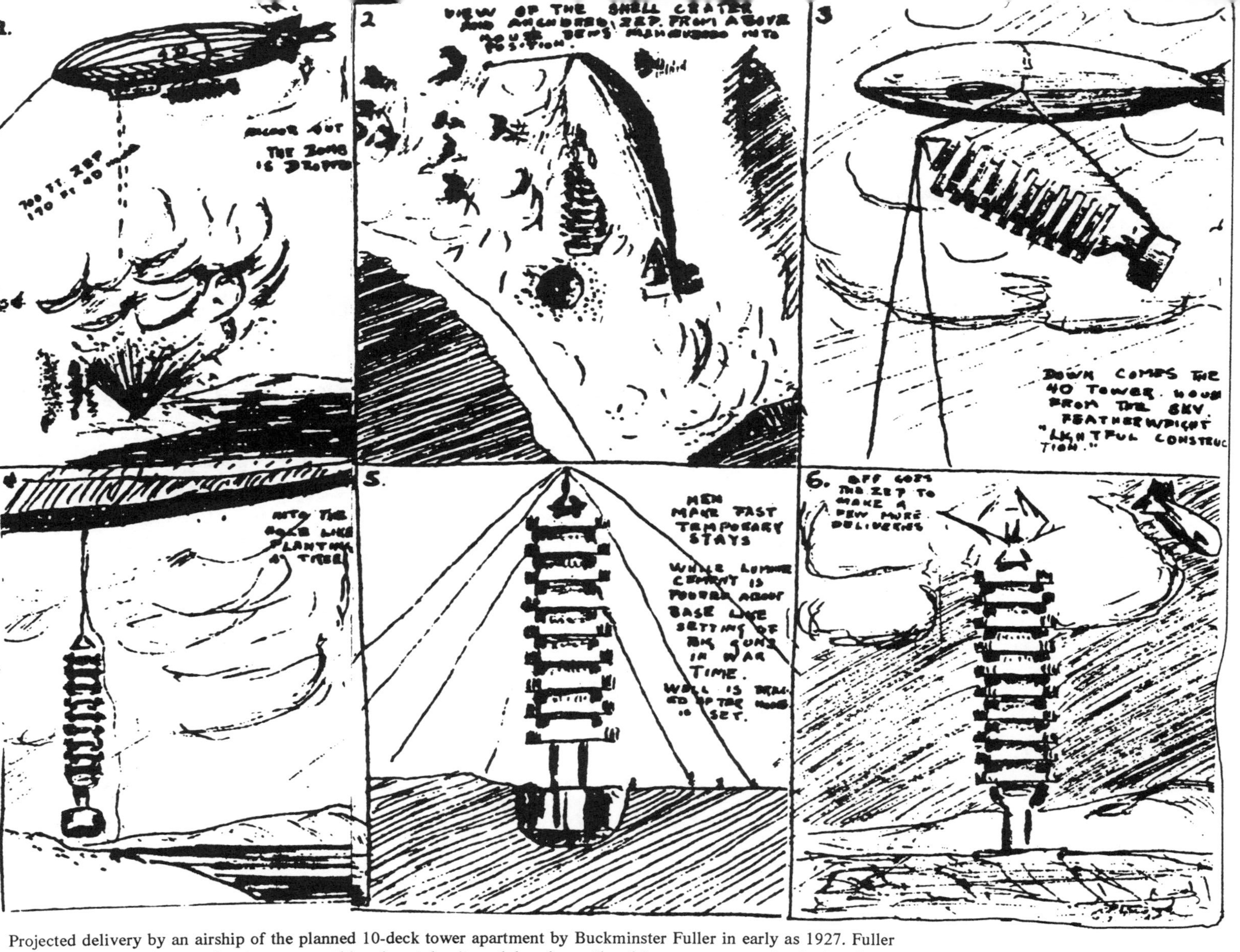

Projected delivery by an airship of the planned 10-deck tower apartment by Buckminster Fuller in early as 1927. Fuller assumed that an airship carrying the tower would anchor at the site and then drop a "bomb" to create a crater for the foundation

The Air Ocean World Town Plan, 1927, showing 10-deck 4D houses, which Fuller sometimes called "stepping stone, world airline maintenance crew environment controls," spotted around the earth in places where nature presented the most hostile conditions. Installation points, inaccessible to man in 1927, included the Arctic Circle, the Alaskan coast, Greenland, the Siberian coast, the central Sahara desert, and the upper Amazon. Great circle air routes, which in 1927 seemed dependent on such maintenance stations, were necessary to link the world's population centers. This drawing pre-dates by five years any map showing great circle air routes. Fuller's original 1927 caption read:

26% of earth's surface is dry land. 85% of all earth's dry land shown is above equator. The whole of the human family could stand on Bermuda. All crowded into England they would have 750 sq. feet each. "United we stand, divided we fall" is correct mentally and spiritually but fallacious physically or materially. Two billion new homes will be required in 80 years.

Feasibility studies showing that it was possible to have controlled environment in inaccessible places gave Fuller what he called a "technical permit" to preview a world integrated by air communications, hence a "one-town world." The "environment control" structures were never built. It took the airlines many years to multiply the range of the airplane to the point where it could "jump the inaccessible places," finally to establish world integrating potentials. Nevertheless, the development of the Air Ocean World Town Plan gave Fuller what he regards as a one-generation advantage in postulating an inherently integrated world, in contrast to the traditional "remotely divided world."

"What will the city of tomorrow be like? Here is the giant plastic, metal and unbreakable glass city of the 21st century. A city of science, of atomic power, of space travel, and of high culture." Thus the editors of *Amazing Stories* magazine described Julian Krupa's back-cover illustration for April 1942. — *Photo by Joe A. Goulait, Smithsonian Institution*

DEAR MR. GRAVITY:

LETTERS ON THE UNIFIED FIELD

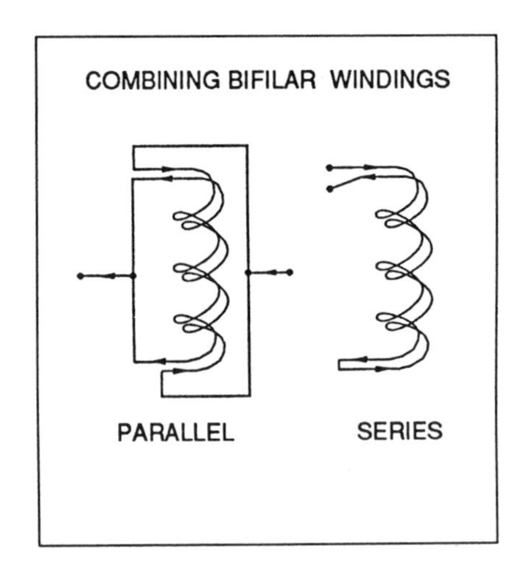

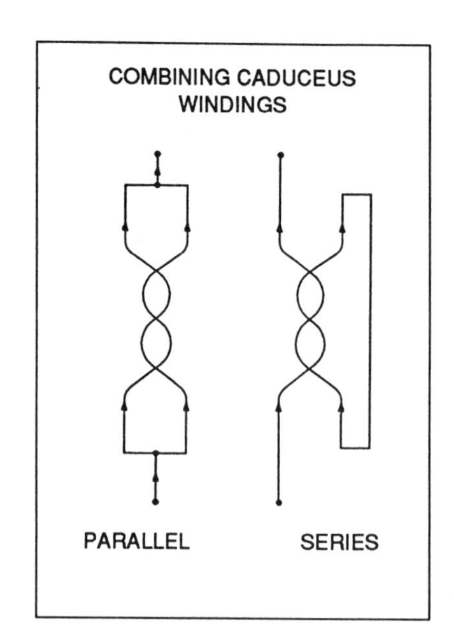

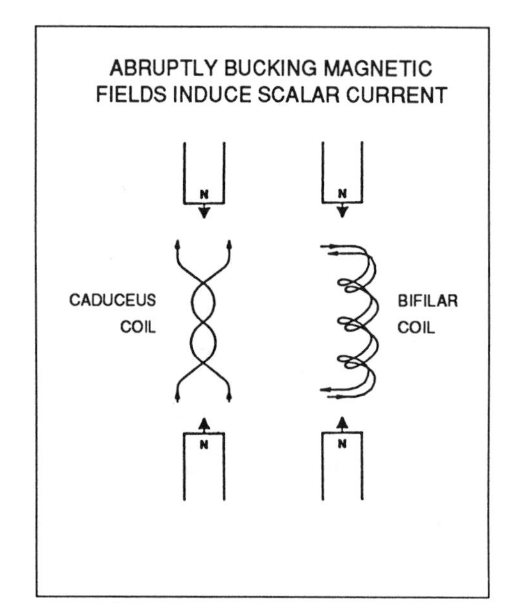

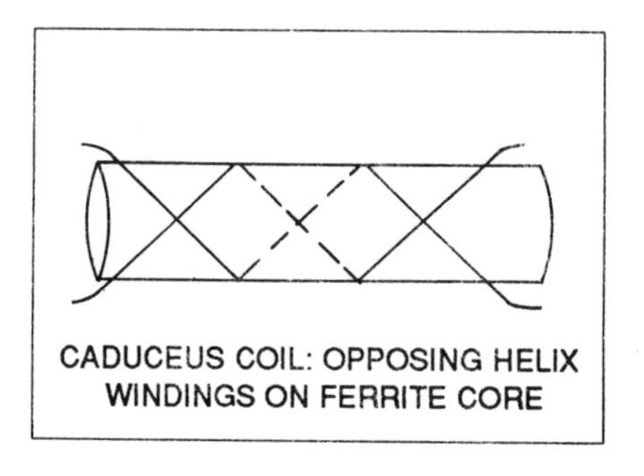

CADUCEUS COIL: OPPOSING HELIX WINDINGS ON FERRITE CORE

Dear Sirs,

I am a man who today noticed that on page 17 of *The Anti Gravity Handbook* it says that you have been able to find out very little about the Dean Drive.

Here is a list of references about the Dean Drive or relevant to it:

United States Patent number 2,886,976: "System for Converting Rotary Motion into Unidirectional Motion," granted to Norman L. Dean. You can buy a printed copy of the patent from the Patent Office (*if* you don't mind the Federal Government knowing that you're *interested* in the Dean Drive.) However, the patent doesn't tell much about how to build a Dean Drive.

John W. Campbell: "The Ultra Feeble Reactions," the December 1958 issue of *Astounding Science Fiction* (Volume LXIV, Number 4, Page 6)

John W. Campbell: "The Space Drive Problem," the June 1960 issue of *Analog Science Fiction/Science Fact* (Volume LXV, Number 4, Pages 83-106)

John W. Campbell: "Report on the Dean Drive," the September 1960 issue of *Analog Science Fiction/Science Fact* (Volume LXVI, Number 1, Pages 4-7)

John W. Campbell: "Instrumentation for the Dean Drive," the November 1960 issue of *Analog Science Fiction/Science Fact* (Volume LXVI, Number 3, Pages 95-99)

William O. Davis: "The Fourth Law of Motion," the May 1962 issue of *Analog Science Fiction/Science Fact* (Volume LXIX, Number 3, Pages 83-104)

William O. Davis, G. H. Stine, E. L. Victory, and S. A. Korff: "Some Aspects of Certain Transient Mechanical Systems," American Physical Society Paper FA10, 1962 Spring Meeting, Washington, D.C. April 23, 1962.

H. Von Shelling: "Stochastic Approach to the Laws of Motion," General Electric Company Report No. 63GL106, Advanced Technology Laboratories, July 1, 1963.

A letter from the inventor, Norman L. Dean, was published in the May 1963 issue of *Analog Science Fiction/Science Fact* (Volume LXXI, Number 3, Pages 4 ff)

More references about the Dean Drive or relating to it:

G.H. Stine and E. L. Victory: a letter published in the September 1963 issue of *Analog Science Fiction/Science Fact* (Volume LXXII, Number 1, Page 4 ff)

Another letter from the inventor, Norman L. Dean, was published in the January 1964 issue of *Analog Science Fiction/Science Fact* (Volume LXXII, Number 5, Page 92 ff)

William O. Davis: "The Huyck Dynamic Systems Research Program: Theoretical Background, Experimental Work, and Some Implications." This is a report of the Huyck Corporation (in about 1961). I don't know whether it was ever published.

William O. Davis: "Some Unusual Implications Inherent in the Huyck Dynamic Systems Research Program." This is a report of the Huyck Corporation (in about 1961). I don't know whether it was ever published.

William O. Davis: "The Energy Transfer Delay Time," Annals of the New York Academy of Sciences, Volume 138, Article 2, Pages 862-863, February 6, 1967)

G. Harry Stine: "Detesters, Phasers, and Dean Drives," the June 1976 issue of *Analog Science Fiction/Science Fact* (Volume XCVI, Number 36 Pages 60-80) Among many other things, G. Harry Stine lists some references, most of which I list here, and says that the last word on the Dean Drive has yet to be written. I don't know whether he has written anything else about it recently.)

Analog Science Fiction/Science Fact is a monthly magazine that is sold on newsstands and in grocery stores, etc. It used to be called *Astounding Science Fiction* until a few decades ago the person who had been the Editor at that time, John W. Campbell, changed the name to *Analog Science Fiction/Science Fact* . John W. Campbell died in (I think) the early 1970s. Also, the inventor, Norman L. Dean, died, but I don't know when. G. Harry Stine is still alive and is publishing articles approximately every other month in *Analog Science Fiction/Science Fact*.

I assume you can write to G. Harry Stine in care of *Analog Science Fiction/Science Fact*, 380 Lexington Avenue, New York, New York, 10017. Perhaps you can also get back issues of *Analog Science Fiction/Science Fact* from some other department at that address, but I don't know.

You can get photocopies of the articles and letters which appeared in *Analog Science Fiction/Science Fact* by buying them from Xerox University Microfilms. Besides microfilms, you can also get photocopied pages. The last I heard (about ten years ago) that company was at 300 North Zeeb Road, Ann Arbor, Michigan, 48106. Or perhaps a library in Chicago has back issues of these magazines.

Pages 30 and 31 of *The Anti Gravity Handbook* give the equation of Albert Einstein as $E=MC^2$." But that is completely incorrect. The correct equation is $E=mc^2$. It should be a small "m" and a small "c." The E is supposed to be a capital letter but the other two letters are supposed to be small letters. See an encyclopedia or a physics book. Or consult the *Handbook of Chemistry and Physics* (which should be in a library.) It will give "m" and "c" as "mass" and "the velocity of light in a vacuum," but it will give entirely different definitions for capital M and capital C. See the section called "Abbreviations and Symbols."

Sincerely,

A. Gravity Researcher

Mr. Gravity answers: *Thankyou Mr. A. Gravity Researcher! We welcome all this information on the Dean Drive and we stand corrected as to the correct symbols for mass and the velocity of light in a vacuum.*

Mr. Gravity welcomes letters from readers and researchers all over the world.

Please write to: Mr. Gravity, c/o Adventures Unlimited Press, Box 22, Stelle, Illinois 60919 USA.

ANTI-GRAVITY PATENTS FOR FUN & PROFIT

"Everything that can be invented has been invented."
—Charles H. Duell
Director of U.S. Patent Office, 1899

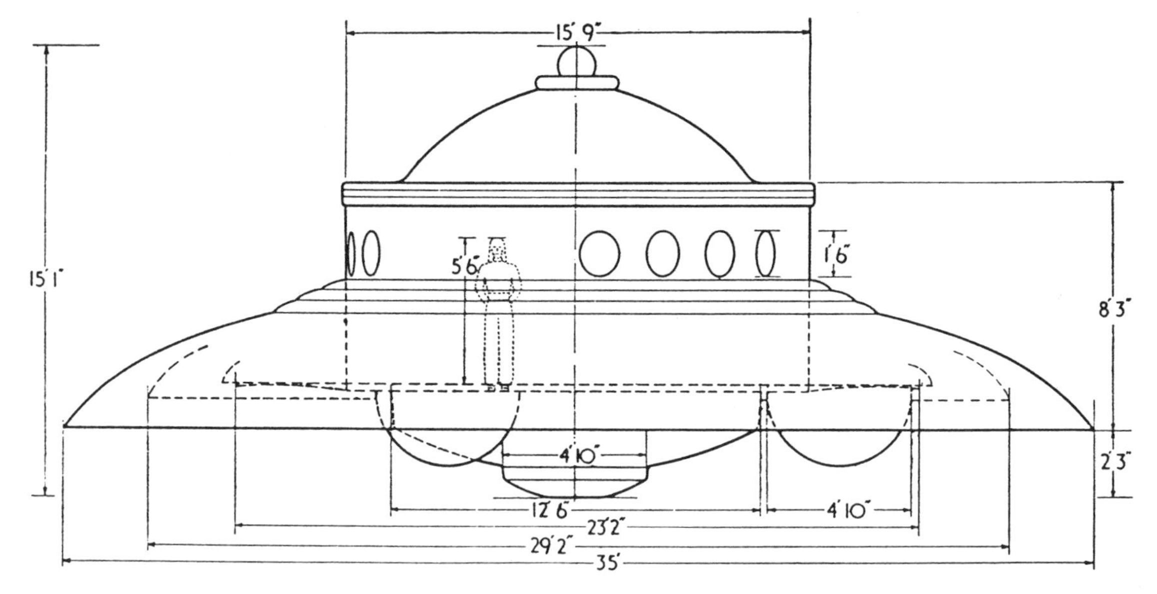

Diagrams courtesy of Leonard G. Cramp, from his book ***Space, Gravity & the Flying Saucer*** (1955).

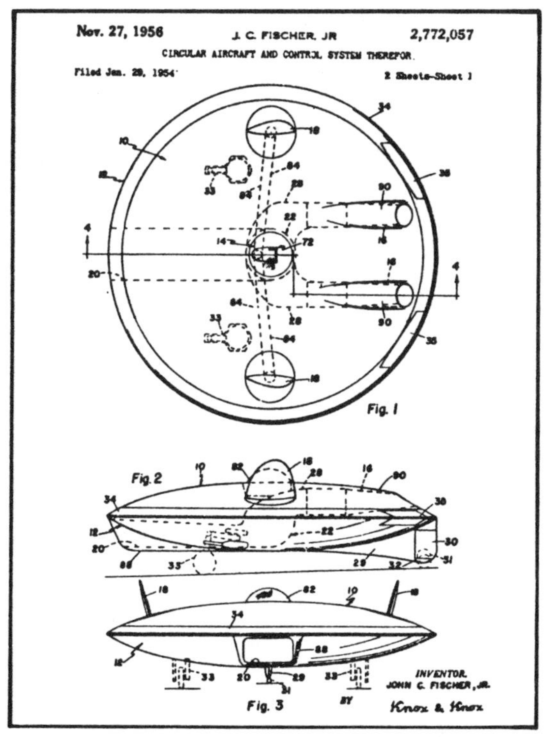

Patented design for John Fischer's rotating circular aircraft.

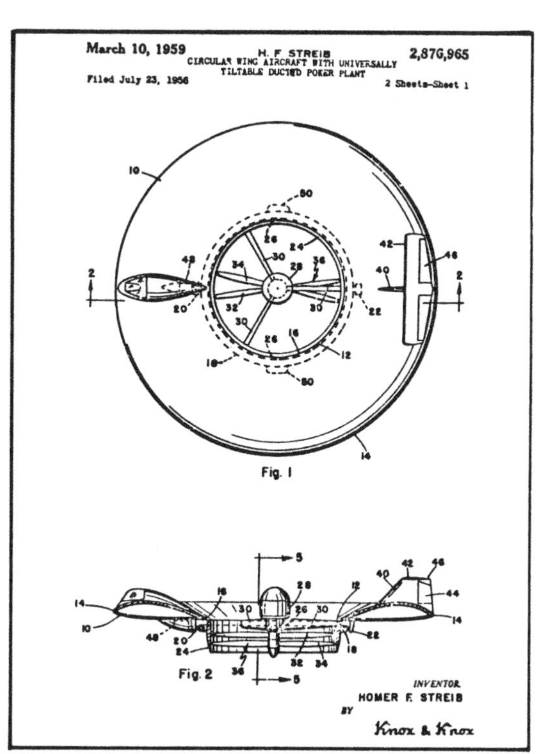

Patented design for Homer Streib's circular wing aircraft, with tiltable ducted power plant, capable of vertical and lateral flight.

Patented design for Homer Streib's circular wing aircraft capable of vertical and lateral flight.

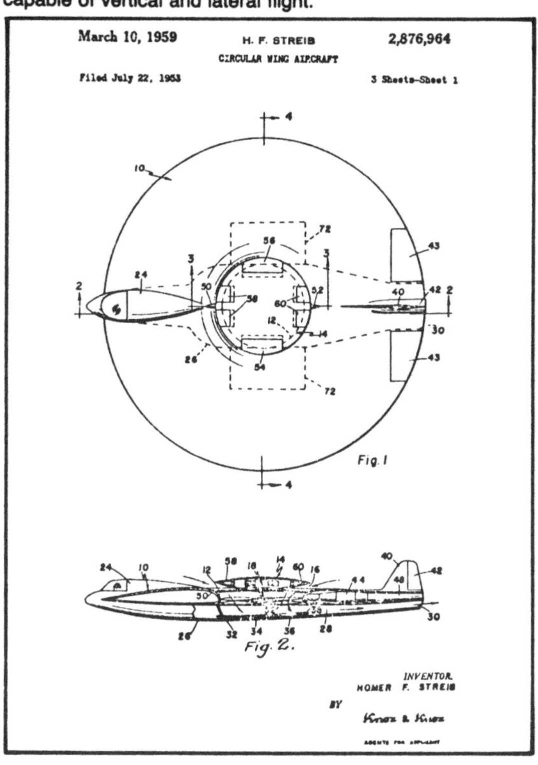

Patented design for Nathan Price's "High Velocity High Altitude V.T.O.L. Aircraft."

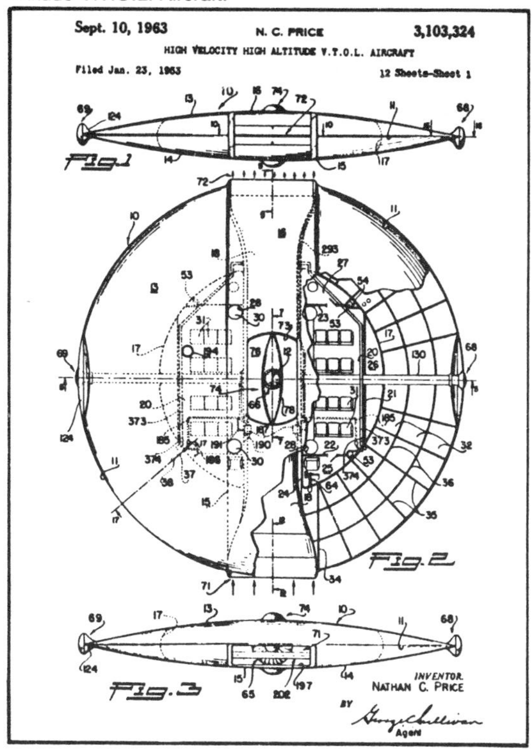

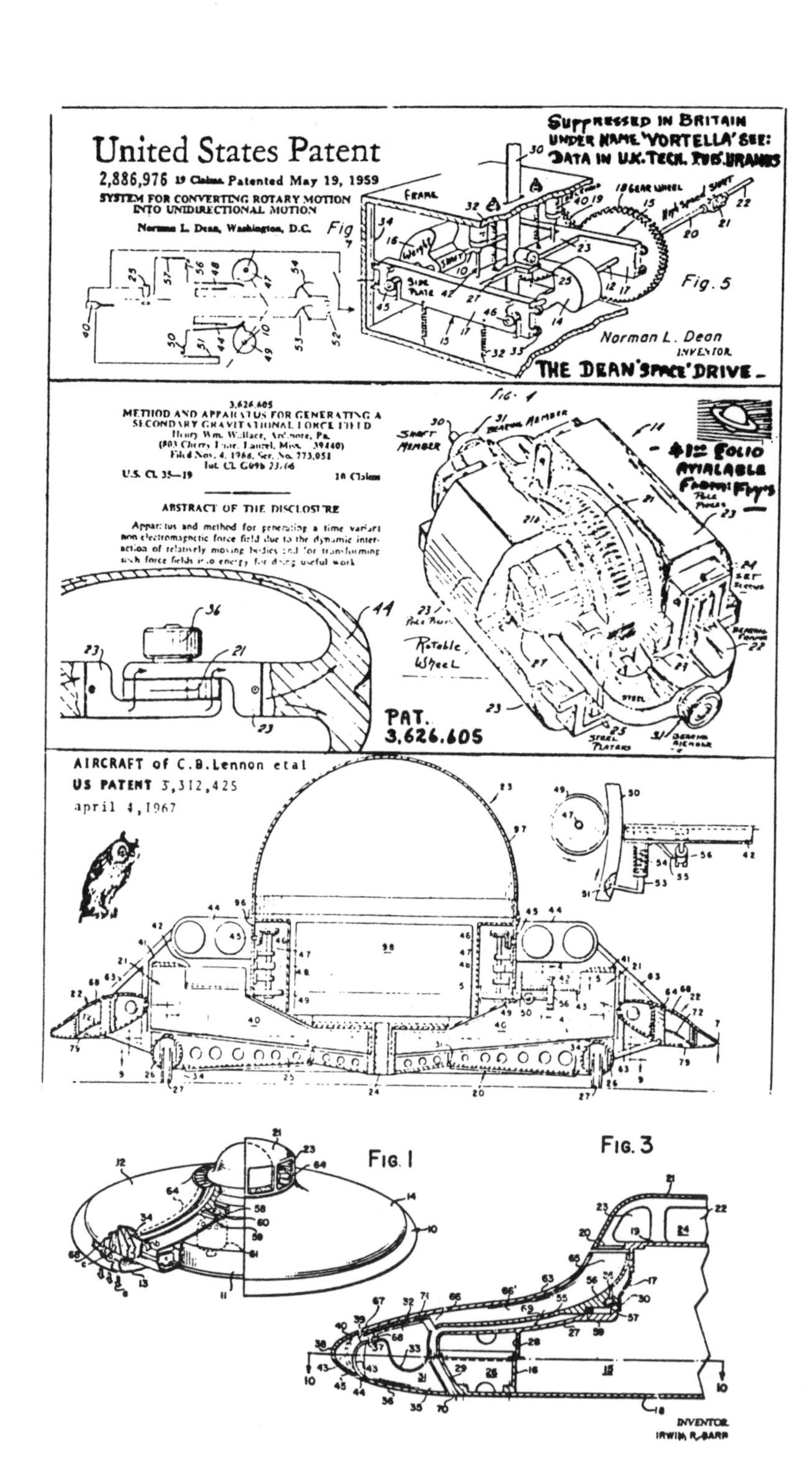

United States Patent
2,886,976 19 Claims Patented May 19, 1959
SYSTEM FOR CONVERTING ROTARY MOTION INTO UNIDIRECTIONAL MOTION
Norman L. Dean, Washington, D.C.
SUPPRESSED IN BRITAIN UNDER NAME 'VORTELLA' SEE: DATA IN U.K. TECH. PUBS. URANUS
Fig. 5
Norman L. Dean
INVENTOR
THE DEAN 'SPACE' DRIVE -
3,626,605
METHOD AND APPARATUS FOR GENERATING A SECONDARY GRAVITATIONAL FORCE FIELD
Henry Wm. Wallace, Ardmore, Pa.
(803 Cherry Lane, Laurel, Miss. 39440)
Filed Nov. 4, 1968, Ser. No. 773,051
Int. Cl. G09b 23/06
U.S. Cl. 35—19
10 Claims
ABSTRACT OF THE DISCLOSURE
Apparatus and method for generating a time variant non electromagnetic force field due to the dynamic interaction of relatively moving bodies and for transforming such force fields into energy for doing useful work.
Rotable Wheel
PAT. 3,626,605
- $1.00 FOLIO AVAILABLE
AIRCRAFT of C.B.Lennon etal
US PATENT 3,312,425
april 4,1967
FIG. 1
FIG. 3
INVENTOR
IRWIN R. BARR

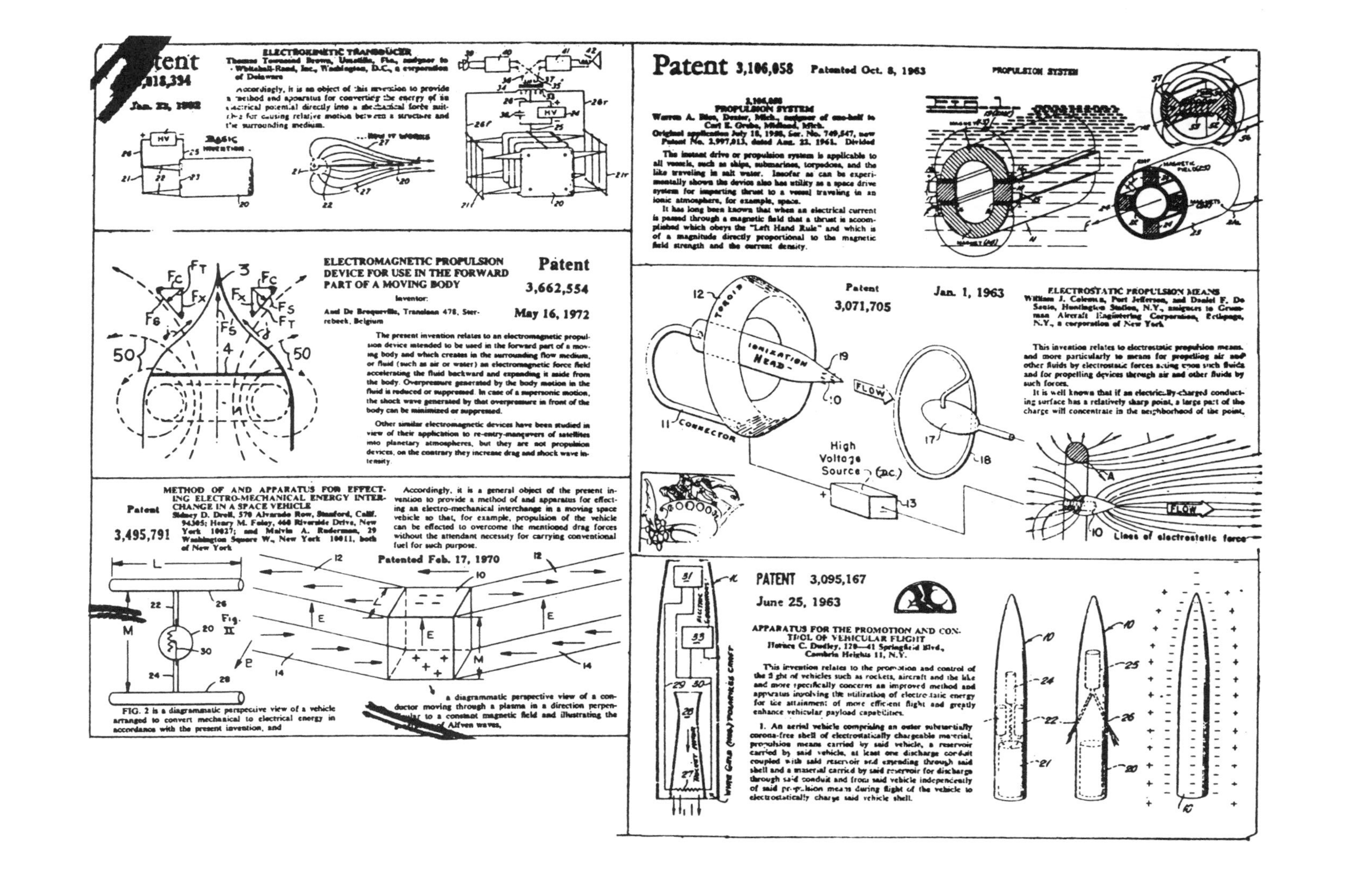

…tent
…,018,334
Jan. 23, 1962

ELECTROKINETIC TRANSDUCER
Thomas Townsend Brown, Umatilla, Fla., assignor to Whitehall-Rand, Inc., Washington, D.C., a corporation of Delaware

Accordingly, it is an object of this invention to provide a method and apparatus for converting the energy of an electrical potential directly into a mechanical force suitable for causing relative motion between a structure and the surrounding medium.

Patent 3,106,058 Patented Oct. 8, 1963

3,106,058
PROPULSION SYSTEM
Warren A. Bien, Dexter, Mich., assignor of one-half to Carl E. Grebe, Midland, Mich.
Original application July 18, 1958, Ser. No. 749,547, now Patent No. 2,997,013, dated Aug. 22, 1961. Divided

The instant drive or propulsion system is applicable to all vessels, such as ships, submarines, torpedoes, and the like traveling in salt water. Insofar as can be experimentally shown the device also has utility as a space drive system for imparting thrust to a vessel traveling in an ionic atmosphere, for example, space.

It has long been known that when an electrical current is passed through a magnetic field that a thrust is accomplished which obeys the "Left Hand Rule" and which is of a magnitude directly proportional to the magnetic field strength and the current density.

ELECTROMAGNETIC PROPULSION DEVICE FOR USE IN THE FORWARD PART OF A MOVING BODY

Patent 3,662,554

May 16, 1972

Inventor:
Axel De Broquerville, Translaan 478, Sterrebeek, Belgium

The present invention relates to an electromagnetic propulsion device intended to be used in the forward part of a moving body and which creates in the surrounding flow medium, or fluid (such as air or water) an electromagnetic force field accelerating the fluid backward and expanding it aside from the body. Overpressure generated by the body motion in the fluid is reduced or suppressed. In case of a supersonic motion, the shock wave generated by that overpressure in front of the body can be minimized or suppressed.

Other similar electromagnetic devices have been studied in view of their application to re-entry-manœuvers of satellites into planetary atmospheres, but they are not propulsion devices, on the contrary they increase drag and shock wave intensity.

Patent 3,071,705 Jan. 1, 1963

ELECTROSTATIC PROPULSION MEANS
William J. Coleman, Port Jefferson, and Daniel F. De Santo, Huntington Station, N.Y., assignors to Grumman Aircraft Engineering Corporation, Bethpage, N.Y., a corporation of New York

This invention relates to electrostatic propulsion means, and more particularly to means for propelling air and other fluids by electrostatic forces acting upon such fluids and for propelling devices through air and other fluids by such forces.

It is well known that if an electrically-charged conducting surface has a relatively sharp point, a large part of the charge will concentrate in the neighborhood of the point,

Patent 3,495,791

METHOD OF AND APPARATUS FOR EFFECTING ELECTRO-MECHANICAL ENERGY INTERCHANGE IN A SPACE VEHICLE
Sidney D. Drell, 570 Alvarado Row, Stanford, Calif. 94305; Henry M. Foley, 460 Riverside Drive, New York 10027; and Malvin A. Ruderman, 29 Washington Square W., New York 10011, both of New York

Accordingly, it is a general object of the present invention to provide a method of and apparatus for effecting an electro-mechanical interchange in a moving space vehicle so that, for example, propulsion of the vehicle can be effected to overcome the mentioned drag forces without the attendant necessity for carrying conventional fuel for such purpose.

Patented Feb. 17, 1970

FIG. 2 is a diagrammatic perspective view of a vehicle arranged to convert mechanical to electrical energy in accordance with the present invention, and

a diagrammatic perspective view of a conductor moving through a plasma in a direction perpendicular to a constant magnetic field and illustrating the … of Alfven waves,

PATENT 3,095,167

June 25, 1963

APPARATUS FOR THE PROMOTION AND CONTROL OF VEHICULAR FLIGHT
Horace C. Dudley, 120—41 Springfield Blvd., Cambria Heights 11, N.Y.

This invention relates to the promotion and control of the flight of vehicles such as rockets, aircraft and the like and more specifically concerns an improved method and apparatus involving the utilization of electrostatic energy for the attainment of more efficient flight and greatly enhance vehicular payload capabilities.

1. An aerial vehicle comprising an outer substantially corona-free shell of electrostatically chargeable material, propulsion means carried by said vehicle, a reservoir carried by said vehicle, at least one discharge conduit coupled with said reservoir and extending through said shell and a material carried by said reservoir for discharge through said conduit and from said vehicle independently of said propulsion means during flight of the vehicle to electrostatically charge said vehicle shell.

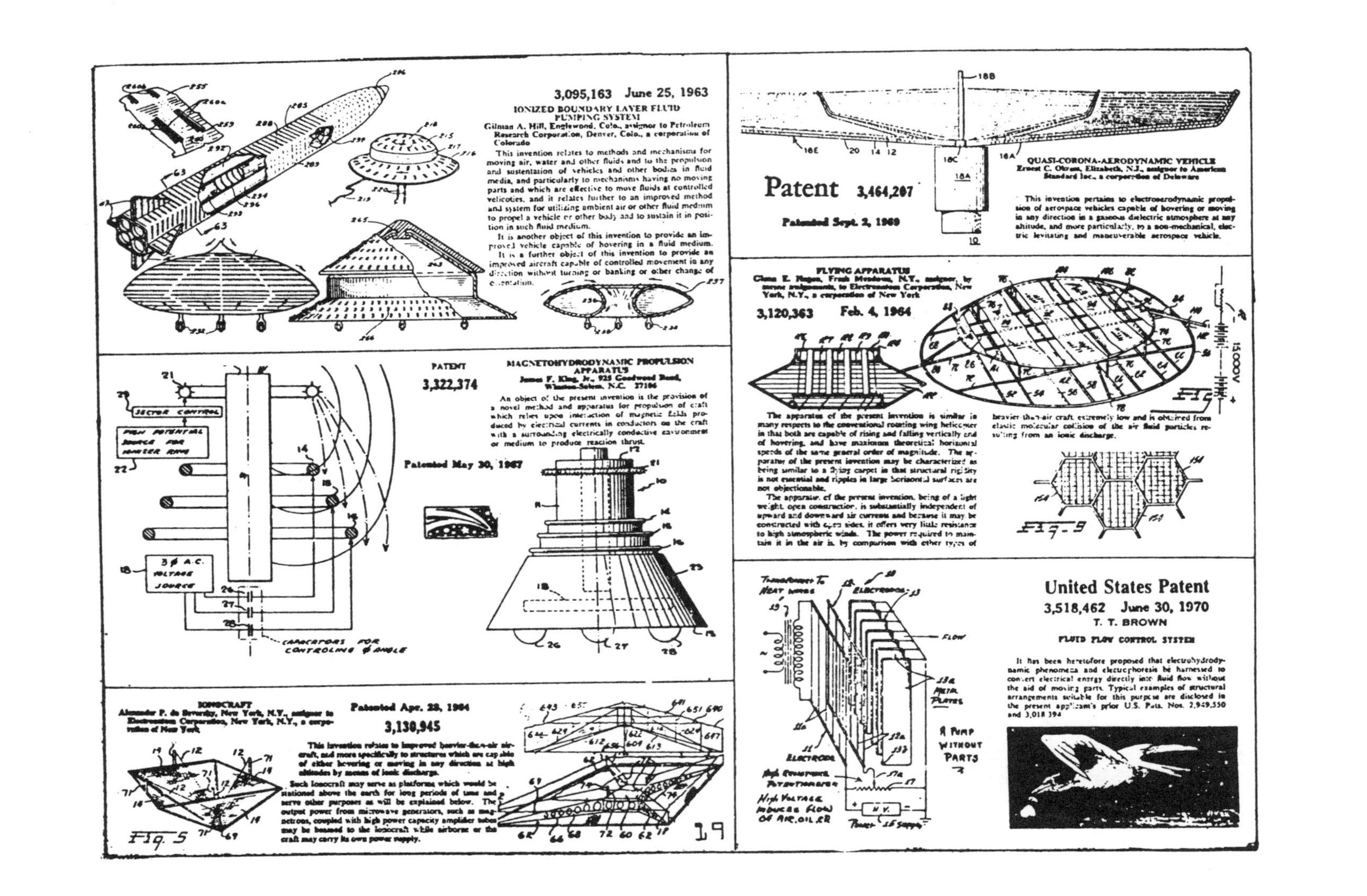

3,095,163 June 25, 1963

IONIZED BOUNDARY LAYER FLUID PUMPING SYSTEM

Gilman A. Hill, Englewood, Colo., assignor to Petroleum Research Corporation, Denver, Colo., a corporation of Colorado

This invention relates to methods and mechanisms for moving air, water and other fluids and to the propulsion and sustentation of vehicles and other bodies in fluid media, and particularly to mechanisms having no moving parts and which are effective to move fluids at controlled velocities, and it relates further to an improved method and system for utilizing ambient air or other fluid medium to propel a vehicle or other body and to sustain it in position in such fluid medium.

It is another object of this invention to provide an improved vehicle capable of hovering in a fluid medium.

It is a further object of this invention to provide an improved aircraft capable of controlled movement in any direction without turning or banking or other change of orientation.

Patent 3,464,207

Patented Sept. 2, 1969

QUASI-CORONA-AERODYNAMIC VEHICLE

Ernest C. Okress, Elizabeth, N.J., assignor to American Standard Inc., a corporation of Delaware

This invention pertains to electroaerodynamic propulsion of aerospace vehicles capable of hovering or moving in any direction in a gaseous dielectric atmosphere at any altitude, and more particularly, to a non-mechanical, electric levitating and maneuverable aerospace vehicle.

PATENT 3,322,374

Patented May 30, 1967

MAGNETOHYDRODYNAMIC PROPULSION APPARATUS

James F. King, Jr., 925 Goodwood Road, Winston-Salem, N.C. 27106

An object of the present invention is the provision of a novel method and apparatus for propulsion of craft which relies upon interaction of magnetic fields produced by electrical currents in conductors on the craft with a surrounding electrically conductive environment or medium to produce reaction thrust.

FLYING APPARATUS

Glenn E. Hagen, Fresh Meadows, N.Y., assignor, by mesne assignments, to Electronatom Corporation, New York, N.Y., a corporation of New York

3,120,363 Feb. 4, 1964

The apparatus of the present invention is similar in many respects to the conventional rotating wing helicopter in that both are capable of rising and falling vertically and of hovering, and have maximum theoretical horizontal speeds of the same general order of magnitude. The apparatus of the present invention may be characterized as being similar to a flying carpet in that structural rigidity is not essential and ripples in large horizontal surfaces are not objectionable.

The apparatus of the present invention, being of a light weight, open construction, is substantially independent of upward and downward air currents and because it may be constructed with open sides, it offers very little resistance to high atmospheric winds. The power required to maintain it in the air is, by comparison with other types of heavier-than-air craft, extremely low and is obtained from elastic molecular collision of the air fluid particles resulting from an ionic discharge.

IONOCRAFT

Alexander P. de Seversky, New York, N.Y., assignor to Electronatom Corporation, New York, N.Y., a corporation of New York

Patented Apr. 28, 1964

3,130,945

This invention relates to improved heavier-than-air aircraft, and more specifically to structures which are capable of either hovering or moving in any direction at high altitudes by means of ionic discharge.

Such Ionocraft may serve as platforms which would be stationed above the earth for long periods of time and serve other purposes as will be explained below. The output power from microwave generators, such as magnetrons, coupled with high power capacity amplifier tubes may be beamed to the ionocraft while airborne or the craft may carry its own power supply.

19

United States Patent

3,518,462 June 30, 1970

T. T. BROWN

FLUID FLOW CONTROL SYSTEM

It has been heretofore proposed that electrohydrodynamic phenomena and electrophoresis be harnessed to convert electrical energy directly into fluid flow without the aid of moving parts. Typical examples of structural arrangements suitable for this purpose are disclosed in the present applicant's prior U.S. Pats. Nos. 2,949,550 and 3,018,394

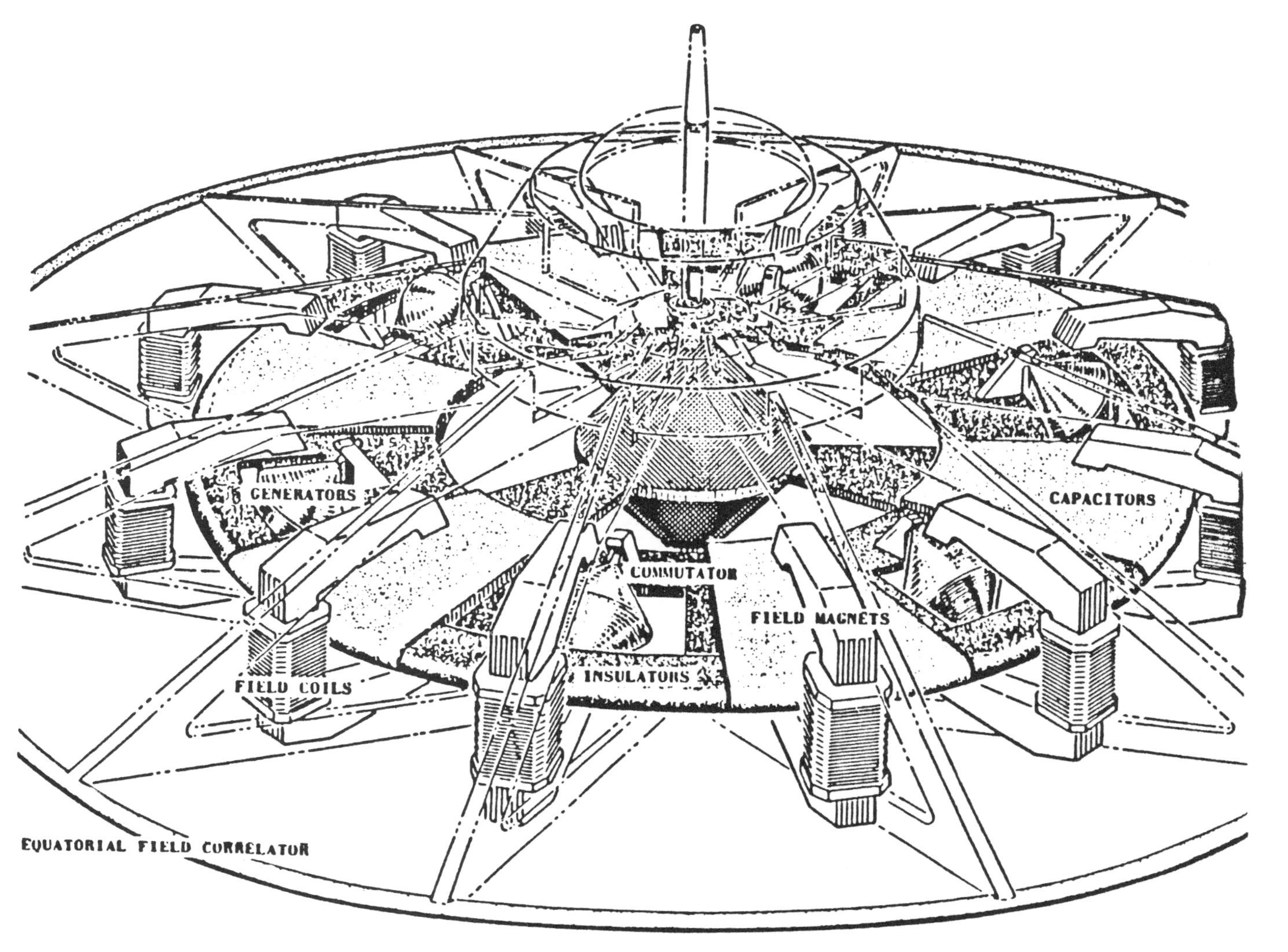

Otis T. Carr's Anti-Gravity Vehicle.

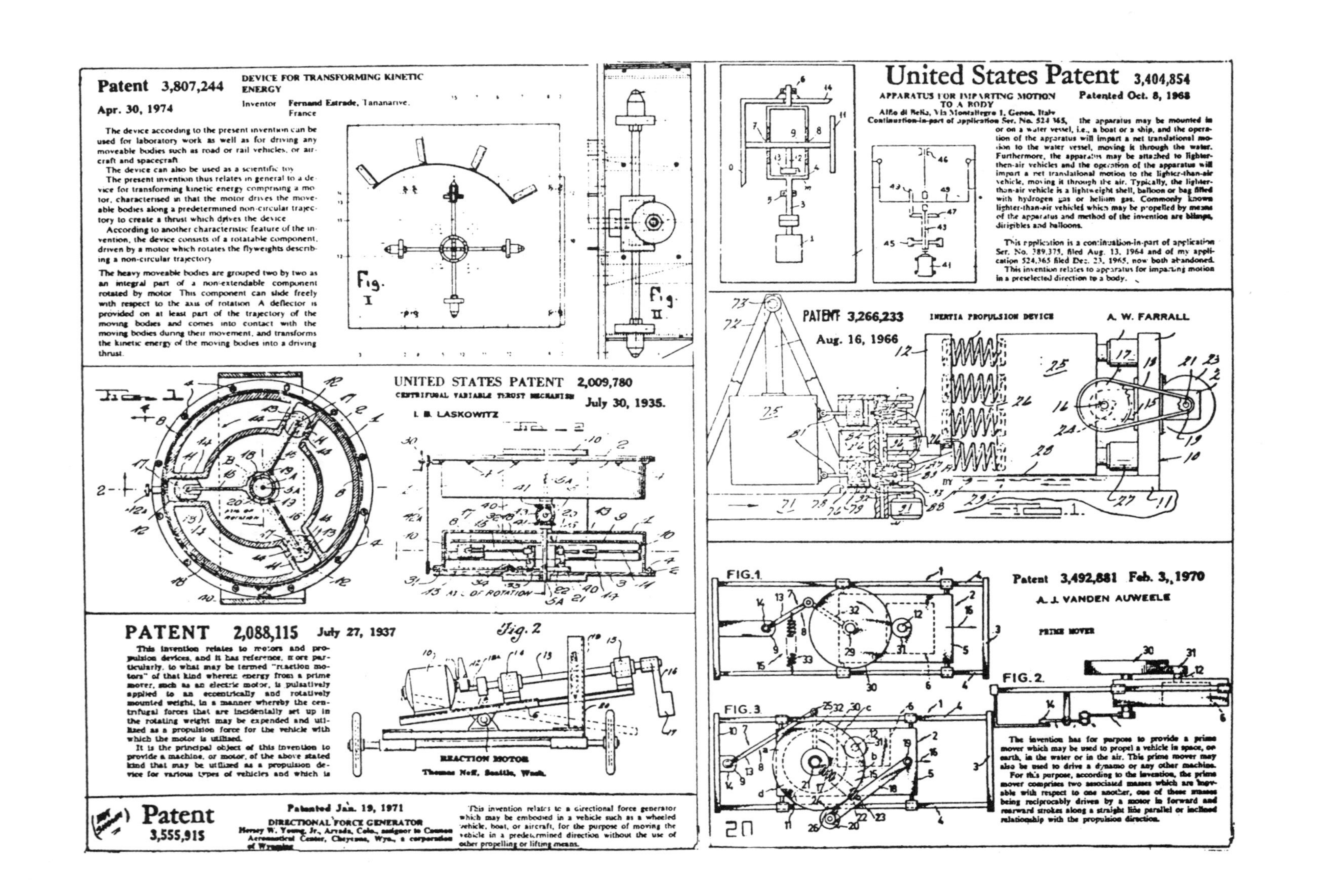

Patent 3,807,244

Apr. 30, 1974

DEVICE FOR TRANSFORMING KINETIC ENERGY

Inventor Fernand Estrade, Tananarive, France

The device according to the present invention can be used for laboratory work as well as for driving any moveable bodies such as road or rail vehicles, or aircraft and spacecraft.

The device can also be used as a scientific toy.

The present invention thus relates in general to a device for transforming kinetic energy comprising a motor, characterised in that the motor drives the moveable bodies along a predetermined non-circular trajectory to create a thrust which drives the device.

According to another characteristic feature of the invention, the device consists of a rotatable component, driven by a motor which rotates the flyweights describing a non-circular trajectory.

The heavy moveable bodies are grouped two by two as an integral part of a non-extendable component rotated by motor. This component can slide freely with respect to the axis of rotation. A deflector is provided on at least part of the trajectory of the moving bodies and comes into contact with the moving bodies during their movement, and transforms the kinetic energy of the moving bodies into a driving thrust.

UNITED STATES PATENT 2,009,780

CENTRIFUGAL VARIABLE THRUST MECHANISM

July 30, 1935.

I. B. LASKOWITZ

PATENT 2,088,115 July 27, 1937

This invention relates to motors and propulsion devices, and it has reference, more particularly, to what may be termed "reaction motors" of that kind wherein energy from a prime mover, such as an electric motor, is pulsatively applied to an eccentrically and rotatively mounted weight, in a manner whereby the centrifugal forces that are incidentally set up in the rotating weight may be expended and utilized as a propulsion force for the vehicle with which the motor is utilized.

It is the principal object of this invention to provide a machine, or motor, of the above stated kind that may be utilized as a propulsion device for various types of vehicles and which is

REACTION MOTOR

Thomas Neff, Seattle, Wash.

Patent 3,555,915

Patented Jan. 19, 1971

DIRECTIONAL FORCE GENERATOR

Hervey W. Young, Jr., Arvada, Colo., assignor to Cosmos Aeronautical Center, Cheyenne, Wyo., a corporation of Wyoming

This invention relates to a directional force generator which may be embodied in a vehicle such as a wheeled vehicle, boat, or aircraft, for the purpose of moving the vehicle in a predetermined direction without the use of other propelling or lifting means.

United States Patent 3,404,854

APPARATUS FOR IMPARTING MOTION TO A BODY

Patented Oct. 8, 1968

Alfio di Bella, Via Montallegro 1, Genoa, Italy

Continuation-in-part of application Ser. No. 524,365,

the apparatus may be mounted in or on a water vessel, i.e., a boat or a ship, and the operation of the apparatus will impart a net translational motion to the water vessel, moving it through the water. Furthermore, the apparatus may be attached to lighter-than-air vehicles and the operation of the apparatus will impart a net translational motion to the lighter-than-air vehicle, moving it through the air. Typically, the lighter-than-air vehicle is a lightweight shell, balloon or bag filled with hydrogen gas or helium gas. Commonly known lighter-than-air vehicles which may be propelled by means of the apparatus and method of the invention are blimps, dirigibles and balloons.

This application is a continuation-in-part of application Ser. No. 389,375, filed Aug. 13, 1964 and of my application 524,365 filed Dec. 23, 1965, now both abandoned.

This invention relates to apparatus for imparting motion in a preselected direction to a body.

PATENT 3,266,233

Aug. 16, 1966

INERTIA PROPULSION DEVICE

A. W. FARRALL

Patent 3,492,881 Feb. 3, 1970

A. J. VANDEN AUWEELE

PRIME MOVER

The invention has for purpose to provide a prime mover which may be used to propel a vehicle in space, on earth, in the water or in the air. This prime mover may also be used to drive a dynamo or any other machine.

For this purpose, according to the invention, the prime mover comprises two associated masses which are movable with respect to one another, one of these masses being reciprocably driven by a motor in forward and rearward strokes along a straight line parallel or inclined relationship with the propulsion direction.

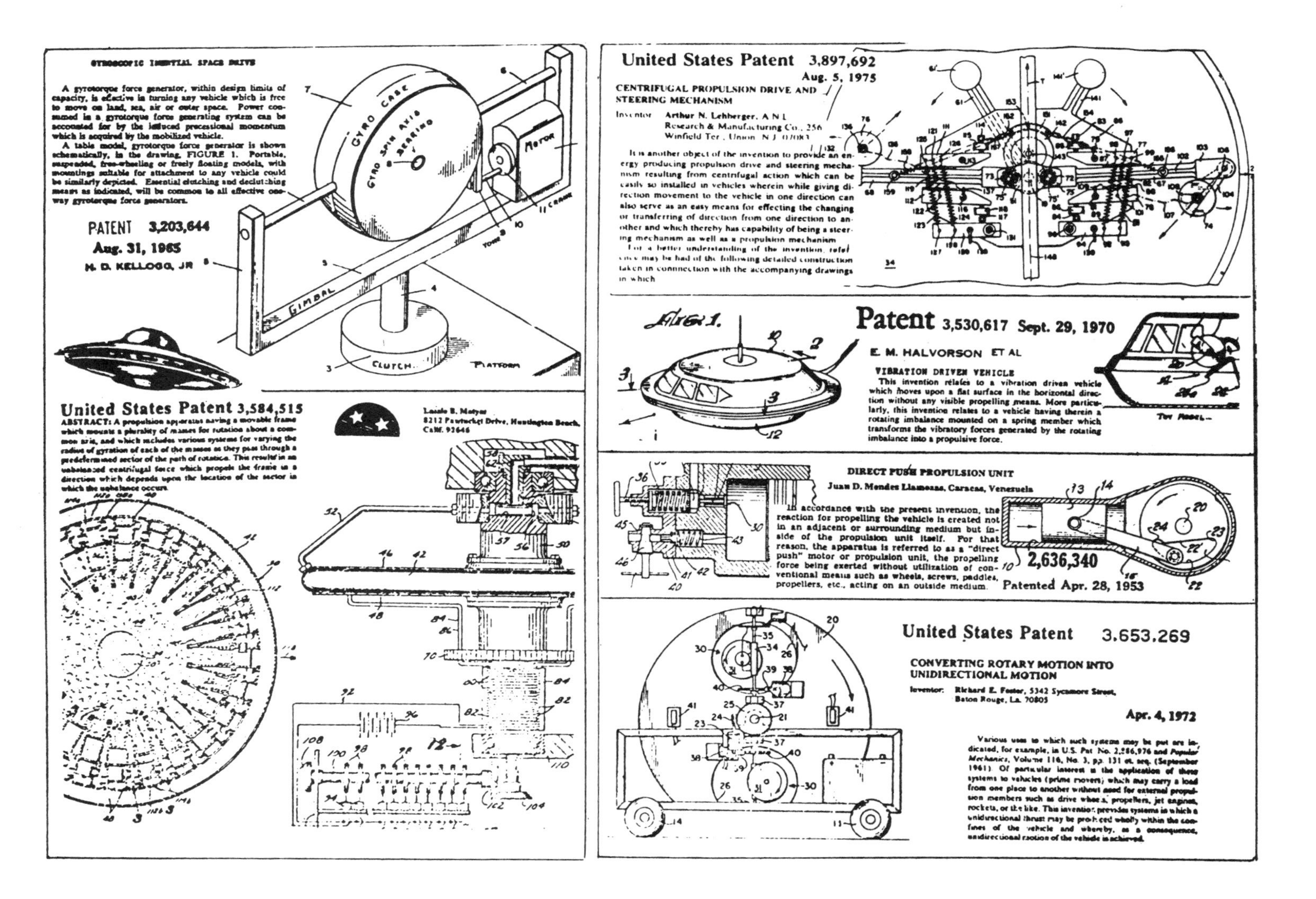

GYROSCOPIC INERTIAL SPACE DRIVE

A gyrotorque force generator, within design limits of capacity, is effective in turning any vehicle which is free to move on land, sea, air or outer space. Power consumed in a gyrotorque force generating system can be accounted for by the induced precessional momentum which is acquired by the mobilized vehicle.

A table model, gyrotorque force generator is shown schematically, in the drawing, FIGURE 1. Portable, suspended, free-wheeling or freely floating models, with mountings suitable for attachment to any vehicle could be similarly depicted. Essential clutching and declutching means as indicated, will be common to all effective one-way gyrotorque force generators.

PATENT 3,203,644

Aug. 31, 1965

H. D. KELLOGG, JR

United States Patent 3,584,515

ABSTRACT: A propulsion apparatus having a movable frame which mounts a plurality of masses for rotation about a common axis, and which includes various systems for varying the radius of gyration of each of the masses as they pass through a predetermined sector of the path of rotation. This results in an unbalanced centrifugal force which propels the frame in a direction which depends upon the location of the sector in which the unbalance occurs.

Laszlo B. Matyas
8212 Pawtucket Drive, Huntington Beach, Calif. 92646

United States Patent 3,897,692

Aug. 5, 1975

CENTRIFUGAL PROPULSION DRIVE AND STEERING MECHANISM

Inventor: Arthur N. Lehberger, A N L Research & Manufacturing Co., 256 Winfield Ter., Union, N.J. 07083

It is another object of the invention to provide an energy producing propulsion drive and steering mechanism resulting from centrifugal action which can be easily so installed in vehicles wherein while giving direction movement to the vehicle in one direction can also serve as an easy means for effecting the changing or transferring of direction from one direction to another and which thereby has capability of being a steering mechanism as well as a propulsion mechanism.

For a better understanding of the invention, reference may be had of the following detailed construction taken in connection with the accompanying drawings in which

Patent 3,530,617 Sept. 29, 1970

E. M. HALVORSON ET AL

VIBRATION DRIVEN VEHICLE

This invention relates to a vibration driven vehicle which moves upon a flat surface in the horizontal direction without any visible propelling means. More particularly, this invention relates to a vehicle having therein a rotating imbalance mounted on a spring member which transforms the vibratory forces generated by the rotating imbalance into a propulsive force.

DIRECT PUSH PROPULSION UNIT

Juan D. Mendes Llameans, Caracas, Venezuela

In accordance with the present invention, the reaction for propelling the vehicle is created not in an adjacent or surrounding medium but inside of the propulsion unit itself. For that reason, the apparatus is referred to as a "direct push" motor or propulsion unit, the propelling force being exerted without utilization of conventional means such as wheels, screws, paddles, propellers, etc., acting on an outside medium.

2,636,340

Patented Apr. 28, 1953

United States Patent 3,653,269

CONVERTING ROTARY MOTION INTO UNIDIRECTIONAL MOTION

Inventor: Richard E. Foster, 5342 Sycamore Street, Baton Rouge, La. 70805

Apr. 4, 1972

Various uses to which such systems may be put are indicated, for example, in U.S. Pat. No. 2,886,976 and *Popular Mechanics*, Volume 116, No. 3, pp. 131 et. seq. (September 1961). Of particular interest is the application of these systems to vehicles (prime movers), which may carry a load from one place to another without need for external propulsion members such as drive wheels, propellers, jet engines, rockets, or the like. This invention provides systems in which a unidirectional thrust may be produced wholly within the confines of the vehicle and whereby, as a consequence, unidirectional motion of the vehicle is achieved.

'The limit to what can be extracted really depends on the ingenuity of the search you request'

Ideas worth millions going for a song

Australian industry is virtually ignoring a vast storehouse of technical information held in the Australian Patents Office – 30 million patents from all over the world, many of which do not apply in Australia and to which access can be had for a mere $10. BOB BEALE reports.

MONAZITE is a little-known Australian mineral export. It is sold overseas for 55 cents a kilogram. Elements that can be refined from it by methods apparently beyond Australian technology bring up to $1,800 a kilogram.

The Australian Patents Office has extracted all the information it holds on monazite – more than 500 patents – as an example of how industry could capitalise on the fact that only three per cent of of the world's annual half-million new patents apply in Australia: the rest are fair game for Australian industry to take up.

Each year Australia exports about 15,000 tonnes of raw monazite, a by-product of sandmining, to France and North America. This amounts to about 60 per cent of the world market for the mineral.

Monazite is not an easy material to process economically but the purified "rare earth" elements extracted from it sell for between $18 and $1,800 a kilogram. These silvery, very reactive metals are increasingly in demand for a wide range of new applications in such things as fluorescent lights, television picture-tube coatings, laser gunsights, and lightweight, high-energy magnets for electric motors.

No monazite refining takes place in Australia, although the CSIRO is researching part of the process and one company is doing a feasibility study for a $60 million plant at Geraldton in Western Australia, which would process 4,000 tonnes of the mineral a year.

The patents office says there is "virtually no consciousness" in Australian industry of the potential to use its extensive records as a technical and commercial reference library.

Occupying 12 floors of Scarborough House at Woden, in Canberra the office's records include 30 million patents from throughout the world – among them copies of more than 500 overseas patents on various aspects of monazite refining. Applications for Australian patent protection cover only six of them, and one of these has been abandoned.

The documents, as required by the patenting process, include full working drawings and complete technical descriptions. While the granting of a patent in no way guarantees that a device or process is technically or commercially feasible, any one of them might prove to be highly lucrative.

A spokesman for the technology information branch of the patent office, Mr Paul Scammell, says there is nothing to stop Australian manufacturers from using any one of the estimated 485,000 new patents each year for which no Australian patent has been sought.

They can freely make and sell the invention here or in any other country where the patent does not apply; or, by improving or modifying it significantly, they can qualify for a new patent for the product around the world.

Mr Scammell said that despite the vast amount of data held on computer by the patent office, including the entire United States patent records, very few Australian companies saw the office as a source of information about products, processes and overseas competitors' research programs.

Out of the hundreds of inquiries from industry each year, only one in 10 was for this type of information, and most of these were from a handful of regulars.

"The patent system is historically regarded as a system of protection," he said. "But it can be used as a science library, and as a valuable planning tool.

"If a company is involved in research and development, a check with the patent office might well show that a particular technical problem has already been solved. There's not much point in spending vast amounts of money to re-invent the wheel.

"About 90 per cent of Japanese companies have a specialist patents officer and they have used the patents system to their advantage extremely well. We're not aware of any Australian company who has developed that approach in the same way."

ALTHOUGH Australian patents are not yet computerised, the patents office does have access to two international computer services – the British Derwent system and the Austrian Inpadoc system – which between them cover most published patent applications from the major industrial countries.

For a small fee the patents office will search these systems for a surprising range of information.

"We can say who is the world's major manufacturer of, say, headlight globes, and whether they are moving out of the market," Mr Scammell said. "The computer will spew out all the applications for patents for headlight globes, by each year, by the inventor's name, or by the country, if you like. The limit to what can be extracted really depends on the ingenuity of the search you request."

For the past three years the office has had information officers in Sydney and Melbourne, with more to come at its network of capital city offices, specifically to handle such requests.

Allowing for relatively short time lags, manufacturers can use the international patent application publishing system effectively to keep tabs on competitors and technical innovations in their field.

Australian law, for example, demands that a patent application be kept secret only for 18 months; it is then published in full and is available to anyone for just $10.

On average, the full patenting process takes about three years from the time of the first application, and protection then lasts from one to 16 years, depending on the inventor annually renewing the patent. A full 16-year patent, including appplication fees, costs about $3,200.

A relatively new arrival on the scene is the petty patent, a simpler, faster and cheaper alternative to cover "fad" inventions (such as toys or games) with a short commercial life span.

Petty patent applications take about six months to be granted, for a fee of just $55 for the first year. To extend the petty patent for the maximum five years costs another $230.

HEATH-ROBINSON HAD NOTHING ON PMM INVENTORS

Science has long scoffed at the idea of a perpetual motion machine, but that hasn't stopped hordes of ingenious hopefuls from coming up with all manner of weird and wonderful devices to circumvent the laws of nature.

In fact, a search of the Australian Patents Office records has detected something of a modern surge of interest in these elusive beasts.

Between 1920 and 1960, Australian patent applications for perpetual motion machines (PMMs) were received at the rate of about one every two years. During the 60s there was an average of one a year.

Now they're positively pouring in at the rate of three a year: but, sadly, those appealing Heath Robinson-style contraptions of yesteryear have succumbed to the age of high-tech.

Now space-age materials and electro-magnetic devices exclusively hold the PMM stage. Still, there has not yet been one successful application, according to the patents office.

Take last year's "energy generator" from a Japanese inventor. He describes it thus: "No primary external heat source is required, but a temperature difference cycle is formed from a cycle of compression and expanding a refrigerant by means of a piston in a heat-insulated cylinder, to apply a temperature change to a magnetic material . . . which is provided in the system, and the internal change of the substance is derived as energy."

Another patent application in 1963 is for a deceptively simple-looking Electromagnetic Radiation Divider, External Vacuum and Anti-gravity Generator, which turns "electromagnetic radiations into cosmic electricity by sending them through right or left-handed, or the mixture of right and left-handed natural quartz crystal . . . or artificially grown piezoelectric crystal . . . [which] has been electrified by pressure." Phew. The Oxford Concise Science Dictionary list three kinds of theoretical perpetual motion.

The first is motion in which a mechanism, once started, would continue indefinitely to perform useful work without being supplied with energy from an outside source – not feasible, says science, because it would contravene the first law of thermodynamics.

The second is motion in which a mechanism extracts heat from a source and converts all of it into some other form of energy – thumbs down again from the second law of thermodynamics.

Perpetual motion of the third kind is a form of motion that continues indefinitely without doing any useful work. The dictionary says (sensibly avoiding any reference to the jaws of politicians) this may be possible on a microscopic scale, but "experience suggests that on a macroscopic scale such a condition cannot be achieved".

For romantics, the only consolation in the whole story is that the patents office staff have an unofficial "Suitcase file".

Entrants for this category usually shuffle in with a suitcase and request a patent for the invention it contains, but are so secretive that they refuse to show it to anyone.

What's the betting they've got some of those nasty old-fashioned mechanical PMMs?

A 1929 perpetual motion machine patent application . . . 'a power unit comprising an element which is permitted to descend under gravitational force for the production of energy and associated means for balancing out the gravitational force on said element whereby it can be restored to its original position by the expenditure of an amount of energy less than that produced by its gravitational descent'.

HEADLINES FROM THE TWILIGHT ZONE: OR "I HAD ELVIS' UFO BABY"

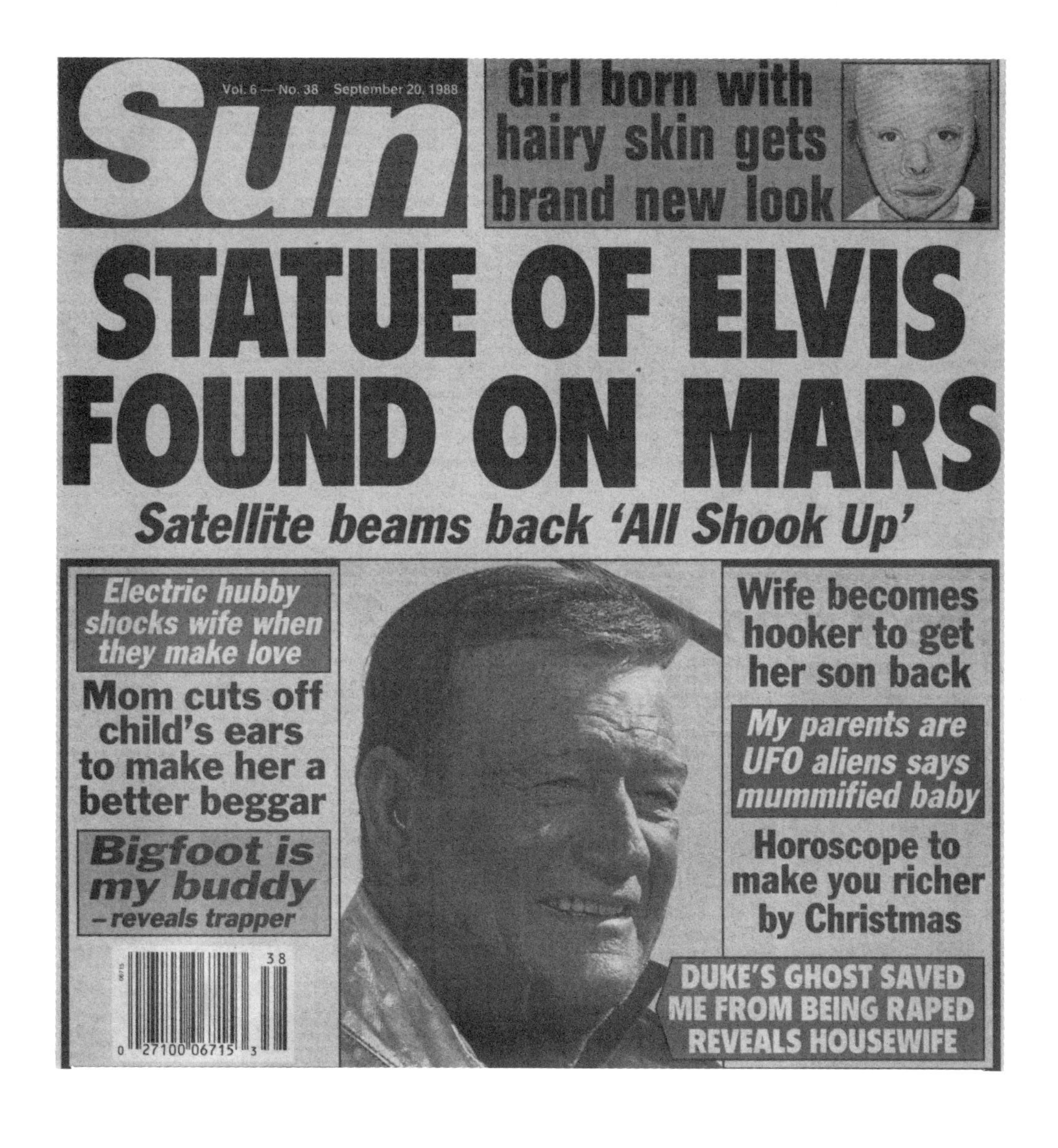

Sun

Vol. 6 — No. 38 September 20, 1988

Girl born with hairy skin gets brand new look

STATUE OF ELVIS FOUND ON MARS

Satellite beams back 'All Shook Up'

Electric hubby shocks wife when they make love

Mom cuts off child's ears to make her a better beggar

Bigfoot is my buddy
—reveals trapper

Wife becomes hooker to get her son back

My parents are UFO aliens says mummified baby

Horoscope to make you richer by Christmas

DUKE'S GHOST SAVED ME FROM BEING RAPED REVEALS HOUSEWIFE

Statue of Elvis found on Mars

• *ALIENS THOUGHT Elvis was one of Earth's gods*

STUNNED SCIENTISTS are trying to make sense out of the most extraordinary discovery in space exploration history — a statue of the rock 'n' roll king that's been found on the surface of Mars!

Photos from a secret Russian space probe show clearly the stone-like image of Elvis Presley standing about 8 feet tall in an isolated desert of the red planet.

Visible

Alien writing still being examined is visible on a plaque beneath the statue.

Also greeting the space probe were the sounds of the popular Elvis song *All Shook Up*, picked up by the radio receiver during its journey.

• *PHOTOS OF the statue were beamed back to Earth*

"The only explanation is the statue was put there by extraterrestrials who think the rock star is some kind of a god on Earth," declares UFOlogist Nikola Stanislaw, who's seen the pictures.

Greeting

"On their visits to our planet they apparently saw the many celebrations of Elvis Presley taking place throughout the world, especially on the anniversary of his death.

"Computer enhancements of the photos show the statue is clearly Elvis.

"It's probably meant as a greeting to establish peaceful relations with us.

"They think Elvis is our lord or king and the writing below the statue may show this.

Fanfare

"The song *All Shook Up*, coming from a device in the statue, is a welcome to our space explorers. The aliens must think this is the proper fanfare because of the music they've heard from Earth," adds Nikola, who is a Hungarian UFO expert.

Nikola learned of the Elvis find through Russian UFO experts working with Soviet scientists studying the high-resolution photos sent back from the top-secret probe.

Cover-up

"The Soviets were going to release the results of the mission until the photos were discovered, and now they're hushing it up," says Nikola.

"But you can't keep a secret of Elvis on Mars very long."

— *FRED SLEEVES*

STUNNED SCIENTISTS...!

Close encounters of the strangest kind

A FLOOD of flying saucer sightings and eerie close encounters with space aliens has led to the formation of a UFO task force.

Skyscan of Worcester, England, is investigating the claims because police have been unable to explain the chain of supernatural events that has shocked an entire nation.

Here are some of the most bizarre incidents:

Four young cattle rustlers told police a spaceship hovered above them in a field in Frodsham, Cheshire. A balloon-like object with a purple glow floated to the ground and two space aliens stepped out.

By GEORGE GLIDDEN

They were humanoid, standing just over five feet tall, and had large, black eyes, small noses, and even tinier mouths. They wore lamps on their heads like miners.

The dumbfounded poachers said the cattle were paralyzed. Using a type of cage, the spacemen seemed to pen one beast and measure it.

Not wanting to be next, the men fled.

A team of top UFOlogists found scorch marks in the field where the craft had landed and traces of radioactivity. Hundreds of citizens dialed police with UFO sightings.

A motorcyclist also spotted a UFO while riding near the field where the poachers saw the ETs.

Vanished

Billy Lowry, 28, said he saw a brightly-lit UFO hovering above some trees.

As he approached the object, it vanished.

Consistent with the pattern of UFO abductions, Lowry claimed he "lost" two hours that evening. He ended up 20 miles from his original destination, lying on the side of the road with strange circular burns on his legs and abdomen.

A 42-year-old banker from London said he was abducted by space aliens as he drove near the town of Little Houghton in Bedfordshire. A brightly-lit spaceship headed straight toward Paul Butterfield's windshield. He blacked out.

Five hours later, Butterfield was found wandering about the countryside with his car keys in his pocket.

The auto was nowhere to be seen.

It was eventually spotted in the middle of a rain-soaked field. The gate was shut and there were no tire marks to show how the car had arrived in the field.

Butterfield, like hundreds of UFO abductees, does not remember what happened to him that evening.

And like many victims he is not willing to discuss the experience.

A nurse set off a full military alert by her report of a UFO. Elda Coretti, 50, was just finishing her shift at 6:35 a.m. when she saw a UFO from a window at Britain's Bath Royal United Hospital.

Big as a double-decker bus, the glowing object soared across the sky, turning from a circular to a cigar shape, she said.

Said Coretti: "I watched it from the window as long as I could. It was yellow-orange. I ran and told my colleagues and then dialed the police."

Added Coretti: "I hadn't been drinking or dabbling in the medicine cabinet. I know what I saw."

Royal Air Force spokesman Ernie Dunsford said the UFO was observed on radar and two jets scrambled to intercept it, but were unsuccessful.

Said Dunsford: "There was certainly something out there."

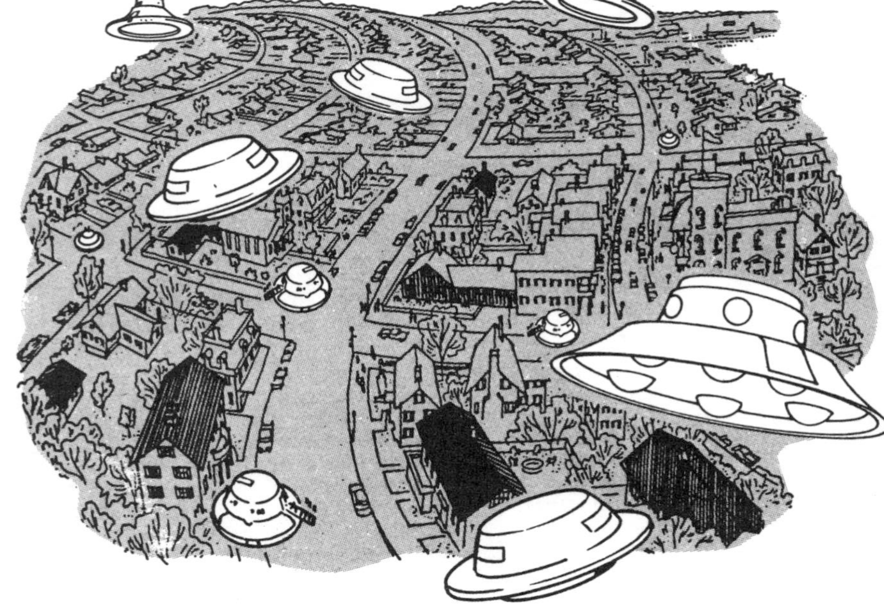

UFO sightings have become almost epidemic in England

June 28, '88/EXAMINER

ENGLISH LEY LINES...

UFO aliens kidnap 1,400 farmers

— astonishing claim

ARE SPACE aliens responsible for the mysterious disappearance of 1,400 people?

That's the question facing police and a UFOlogist in Sha Tin, China, where 1,432 residents have vanished from farms and villages in the past three years, half of them between the ages of seven and 16.

UFOlogist Wong Watin, 52, sees a connection between the disappearances and a rash of UFO sightings.

Wong saw a brightly-lit UFO in December, 1987, about 80 feet long, 60 feet wide, and shaped like a watermelon. It apparently had no windows or any means for the ETs to view the outside.

Zoomed

The UFO stood in a field about 50 feet from the bushes where Wong, a herbalist by profession, routinely picked berries. He crouched low in a ditch and watched in amazement as the UFO rose silently from the ground and zoomed into the sky.

At first Wong suspected the UFO was a Soviet spy plane, but when hundreds of peasants began disappearing, his suspicions turned to ETs.

Wong is now convinced space aliens are kidnapping human beings and performing strange genetic experiments.

Four dazed teens, who had apparently blacked out for several hours, were discovered wandering the countryside with circular burn marks on their arms and legs.

One of the teens, a 16-year-old girl, claimed gray-skinned ETs with big heads and large black eyes stole her unborn baby, according to police.

And a schoolteacher reported that she saw a UFO moving slowly a few feet above the ground toward a pupil on his way home. She grabbed the child and they hid in some bushes.

But people in Sha Tin are panic-stricken, Wong says. They no longer leave home at night and they walk in groups of 10 to 20 during the day.

One farmer said he saw a UFO in his pasture. After it zoomed into the sky, he claimed a horse and a cow were missing.

Mutilated

Another farmer, Cheng-Wu, 55, said he found four of his best cows mutilated. One of the cows had an incision in its head and its brains were missing.

Said Cheng: "There were no signs of poachers. The fields were very muddy and they would have left tracks. This is all a mystery to me."

UFOlogist Watin believes space beings experimented on the cows, lifting them into their spacecraft with a strange beam of light and placing them back in the field when they were finished.

— GREGORY BANGS

SPACESHIP SIGHTINGS in the Orient: Is there a connection with vanishing villagers?

KIDNAPED FARMERS!

SUNDAY EXPRESS, London, England - Oct. 23, 1988

Garden-shed genius heads for the stars

by MICHAEL SHANAHAN

DAY trips to Australia and weekend jaunts in the Milky Way could become a reality, thanks to the brainpower of a Scottish inventor.

Sandy Kidd's discovery, which is set to revolutionise travel, is already sending shock waves through the scientific establishment.

One of Britain's top physicists described it "mind-boggling."

Mr Kidd's work, researched in his garden shed, will make science fiction writers' dreams come true. Trips to Mars will take 34 hours and the journey from London to Sydney will be reduced to a matter of minutes.

The 51-year-old former apprentice toolmaker's development of Gyroscopic Propulsion has also stunned academics because it challenges Isaac Newton's Law of Motion.

Kidd: Breakthrough

He was worked out that, by setting gyroscopes at particular angles, a lifting force that defies gravity is produced.

Mr Kidd, who worked for five years on his brainchild at his Dundee home, is now moving to the heart of the space industry in California where a massive investment programme is already under way.

Dr Harold Aspden, senior visiting research fellow at Southampton University, has seen the results of early tests. "Scientifically speaking it is a bombshell," he says. I would not have believed this if I had not seen it with my own eyes.

"It will totally revolutionise the travel industry. Taken to the ultimate, we will have planes without jet engines and helicopters without rotor blades."

Mr Kidd is being financed by an Australian research company. A spokesman said: "We are on to something really big. The next stage is to power up Sandy's device in California with the prospect of building a full-scale vehicle at the end of the day.

"Money is no object, but we are determined that his work will not get out.

At Imperial College, London, Professor Eric Laithwaite, who has followed Mr Kidd's experiments, said:

"I have always been convinced it could be done . . . and I like to see someone defeat the system. He may be a long time perfecting it but I am sure he will succeed."

Mr Kidd made the final breakthrough in his work highlighted in the Sunday Express last year, about four weeks ago in a laboratory in Melbourne.

"There was just one thing I couldn't understand," said the former RAF radar technician. "I had worked round the problem until that day when it dawned on me. If I could figure it out why hasn't somebody else?"

Is your co-worker a space alien?

It appears that the supermarket tabloids have not backed quite so far away from their once fearless reporting on UFOs and space aliens as a recent report in this space may have led you to believe.

We have come into possession of a story from the National Enquirer in which two "experts"—Brad Steiger, who is identified as a renowned UFO investigator and author, and Dr. Thomas Easton, who is identified as a theoretical biologist and futurist—list 10 signs to watch for if you suspect that some of your apparently human co-workers are in fact aliens from space:

1. Odd or mismatched clothes. Space aliens often do not fully understand styles. So they wear combinations that are bizarre or in bad taste, such as a tuxedo jacket with blue jeans or sneakers.
2. Strange diet or unusual eating habits. Aliens might eat French fries with a spoon or consume large amounts of pills.
3. Bizarre sense of humor. Misunderstanding earthly humor, they may laugh during a company training film or tell jokes that make no sense.
4. Takes frequent sick days. An alien may need extra time to rejuvenate its energy.
5. Keeps a written or tape-recorded diary. Aliens are constantly gathering information.
6. Misuses everyday items. An alien may paint its nails with correction fluid.
7. Constant questioning about customs of co-workers. Aliens may ask questions that seem stupid, such as why do so many Americans picnic on the Fourth of July?
8. Secretive about personal lifestyle and home. An alien won't discuss domestic details or talk about what it does at night or on weekends.
9. Frequently talks to itself. An alien may be practicing our language.
10. Displays a change of mood or physical reaction when near certain high-tech hardware. An alien may, for example, experience a mood change when a microwave oven is turned on.

Don't look now, but we appear to be surrounded!

Clarence Petersen

STRANGE COWORKERS...

Anti-gravity gyroscope raises eyebrows

Washington Post

Claims of inventing an anti-gravity machine are common on the fringes of modern physics and none has ever been taken seriously. Until now.

One of the world's most respected physics journals, *Physical Review Letters*, has published a report by two Japanese scientists who claim that when they spun a special gyroscope at speeds between 3,000 and 13,000 rpm, it lost a tiny amount of weight. The faster it spun, the more weight it lost.

Weight is a measure of gravity's effect and although the weight loss was small – between 20 and 60 millionths of the gyroscope's 11.3 ounces – any provable loss would appear to violate the known laws of physics. Other physicists who have seen the report suspect it is almost certainly wrong, but no one, including several senior physicists who reviewed the research before the journal agreed to publish it, has yet discovered a source of any error.

Unlike most claims of extraordinary phenomena, this one was published with full technical details so that others may try to reproduce the results. The scientists, Hideo Hayasaka and Sakae Takeuchi of Tohoko University in Sendai, **appear to have designed their experiments to rule out many potential sources of error, other physicists say.**

ARKANSAS DEMOCRAT, Little Rock, AR – Jan. 2, 1990

Gyroscope test possibly defies gravity

By **WILLIAM J. BROAD**
New York Times News Service

Japanese scientists have reported that small gyroscopes lose weight when spun under certain conditions, apparently in defiance of gravity. If proved correct, the finding would mark a stunning scientific advance, but experts said they doubted that it would survive intense scrutiny.

A systematic way to negate gravitation, the attraction between all masses and particles of matter in the universe, has eluded scientists since the principles of the force were first elucidated by Isaac Newton in the 17th century.

The anti-gravity work is reported in the Dec. 18 issue of Physical Review Letters, which is regarded by experts as one of the world's leading journals of physics and allied fields. Its articles are rigorously reviewed by other scientists before being accepted for publication, and it rejects far more than it accepts.

Experts who have seen the report said that it seemed to be based on sound research and appeared to have no obvious sources of experimental error, but they cautioned that other seemingly reliable reports have collapsed under close examination.

The work was performed by Hideo Hayasaka and Sakae Takeuchi of the engineering faculty at Tohoku University in Sendai, Japan.

Unlike the exaggerated claims made for low-temperature, or "cold," nuclear fusion this year, the current results are presented with scientific understatement. The authors do not claim to have defied gravity, but simply say their results "cannot be explained by the usual theories."

"It's an astounding claim," said Robert L. Park, a professor of physics at the University of Maryland who is director of the Washington office of the American Physical Society, which publishes Physical Review Letters. "It would be revolutionary if true. But it's almost certainly wrong. Almost all extraordinary claims are wrong."

If substantiated by further tests, the finding could have a profound influence on physics and the study of the universe and perhaps in the making of practical anti-gravity devices.

The experiment looked at weight changes in spinning mechanical gyroscopes whose rotors weighed 140 and 176 grams, or 5 and and 6.3 ounces. When the gyroscopes were spun clockwise, as viewed from above, the researchers found no change in their weight. But when spun counterclockwise, they appeared to lose weight.

ANTI-GRAVITY...

Amazing satellite photos confirm . . .

MIND-BOGGLING MYSTERY: A vintage plane is sitting on the moon, says the head of a Soviet lunar probe. "We have the photos to prove it."

WORLD WAR 2 BOMBER FOUND ON THE MOON!

Russians: Perfectly preserved warplane in crater

Soviet scientists claim to have found a vintage American warplane on the moon!

Dr. Stanislav Makeyev said satellite photos of the World War 2 bomber indicate that it has sustained some damage from meteorites but is still intact.

Air Force insignias are visible on the wings and fuselage, he added. It also appears that the entire plane has a greenish tinge — as if it was covered with algae from the sea.

"We have absolutely no explanation for this and doubt seriously that the Americans have one either," the expert told reporters. "We can speculate that the plane was hijacked by extraterrestrials and taken to the moon.

"But again, that is only speculation. We probably will never know how the plane got there or why."

U.S. officials declined to comment on the Soviet photos and one flatly called the idea of finding an airplane on the moon "preposterous."

But Dr. Makeyev insisted that the satellite's high resolution pictures are genuine — and clearly show the huge bomber at the bottom of a crater in a previously uncharted quadrant of the moon.

Whether or not the Soviets plan to continue their investigation with a mechanical moon probe or actual manned expedition remains to be seen. But Wilhelm Greder, chairman of the Swiss UFO group CONTACT, urged the U.S. and the U.S.S.R. to pool their resources "to get to the bottom of this thing now."

"We have the questions, now we need the answers," Greder told reporters in Geneva. "The fact that the plane is covered with algae tells me there might be a Devil's Triangle connection. This plane could be the proof we need to show that extraterrestrials have a base on the ocean floor and that they have been snatching boats, planes and people in the Triangle for years."

The superpowers need to find out what is behind this discovery.

WHITE RECTANGLE indicates area on lunar surface mapped by the Soviet probe.

Did the FAA have it on radar?

GIGANTIC UFO THAT SHOCKED THE WORLD!

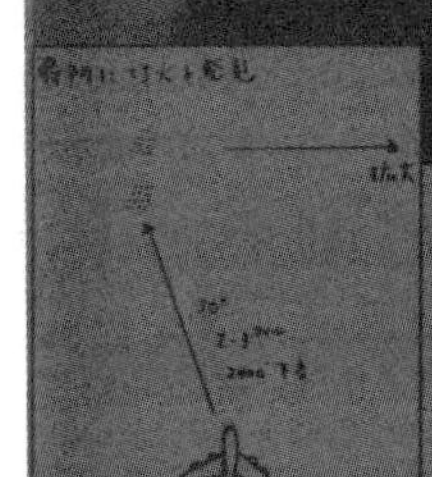

JET pilot Kenju Terauchi's sketch shows where the UFO lights first appeared 8 miles in front of his plane.

By DICK DONOVAN

The U.S. government at first confirmed, then mysteriously denied that a huge UFO, described as bigger than two aircraft carriers, was tracked on radar as it played a bizarre cat-and-mouse game — with a Boeing 747 jet!

FAA air traffic controllers had flatly stated they had tracked the gigantic spacecraft for more than 32 minutes as it followed a Japan Air Lines cargo flight bound for Anchorage, Alaska.

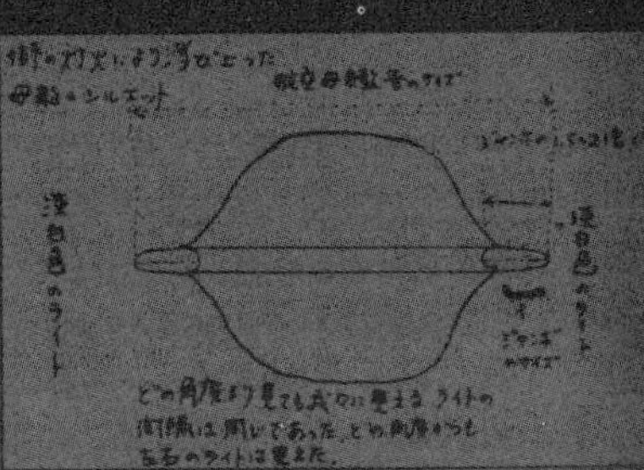

JUMBO jet was dwarfed by huge walnut shaped UFO. This sketch by pilot Terauchi shows the relative sizes of the spacecraft and his 747. The plane is the tiny black shape at right.

Then, in a surprise official flip-flop, an agency spokesman said the eerie blip that zipped helter-skelter across the radar scope was merely a duplicate image of the 747.

That bit of mumbo jumbo, however, flies in the face of eyewitness accounts of the jet's three-man crew that had been kept secret for six weeks — until a crewmember leaked the story to the press.

According to a vivid and detailed account of the incident by veteran JAL pilot Kenju Terauchi, his 747 was flying in clear skies at 35,000 feet and cruising at 525 knots when three walnut-shaped UFOs, the huge one and two smaller ones, streaked out of the heavens.

"We could all see the UFOs very clearly," the 47-year-old pilot said. "One was very large . . . two times bigger than an aircraft carrier. It dwarfed our 747.

"The UFOs were flying parallel and then suddenly approached very close. They moved with amazing speed."

It was not from this earth, says pilot of jumbo jet after close encounter off Alaska

According to initial FAA flight control reports, the UFOs dogged the 747 for at least 32 minutes.

Terauchi, however, said the ships followed him for 400 miles.

Terauchi, whose flying career spans 29 years, said he and his crew tried to escape the UFOs by following FAA instructions to descend 4,000 feet and make several evasive maneuvers.

PILOT'S sketch of plane's radar shows how UFO appeared (blip, lower left).

But the 747 jet was no match for the maneuvering ability of the spacecraft.

"They were still following us," Terauchi said, and FAA radar at that time confirmed that at least one of the UFOs remained nearby.

FAA investigators questioned the 747's crew in Anchorage and said they are "normal, rational, professional" people with no drug or alcohol problems.

The 747 was carrying a cargo of wine bound for Tokyo from Paris when the UFOs appeared eight miles ahead.

Terauchi radioed that the lights he saw were yellow, amber and green, but not red, which is the international color for aircraft beacons.

Paul Steucke, the FAA spokesman in Anchorage, said his agency is continuing its investigation into the incident and that radar tapes and the recorded radio messages are being sent to Washington.

But Terauchi said his only conclusion is that the three UFOs he saw on that November 17 flight were not from earth.

"It was incredible," he said.

Driver aims at ferry but misses boat

Widow Jessica Rawlings missed the boat and narrowly avoided a disaster when she drove her car down a ferry slip — and straight into the water.

The 74-year-old Poole, England, woman was rescued by courageous bystanders who waded into the icy water in 60 m.p.h. winds to pull her from her car.

"It's entirely my own fault," said the flustered oldster.

"I thought the ferry was there, but when I drove down the ramp it wasn't. The next thing I knew I was in the water."

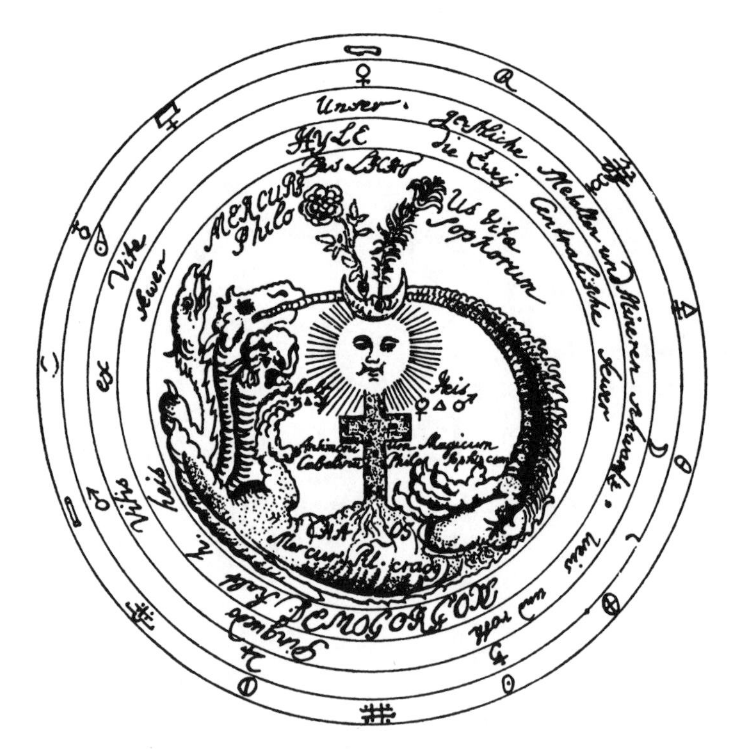

A medieval Rosicrucian alchemical diagram in which mercury plays an important part. The symbols and concentric circles seem to correspond to various elements as well as the electromagnetic fields of an atom. According to the legend, "Whoever deciphers the alchemical riddle will be able to journey through lands otherwise inaccessible."

At 3:00 in the afternoon on March 8, 1964, Harry Hauxler of West Germany took this photo of discoid craft through the window of a train near Oberwesel. Note the whirling dark vortex beneath the rising craft. This craft may well be a Vimana with a mercury vortex propulsion system. Photo courtesy of UFO Photo Archives, Tuscon, Arizona.

THE ANTI-GRAVITY FILE

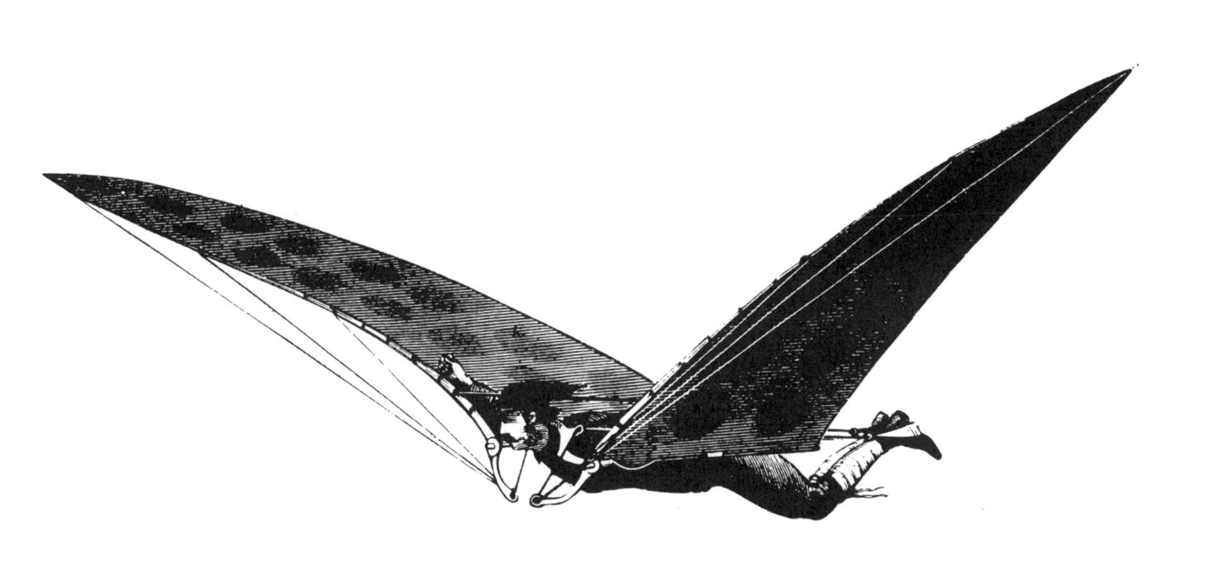

Two photos of an alleged anti-gravity craft photographed in the Caucausus Mountains in Azerbaijan sometime around 1983. Speculation about the craft includes that it was part of a long-secret Soviet space program, a prop for a science-fiction movie, or an early anti-gravity craft, possibly from World War II or before. Note the similar look of this craft to the Lunar Module used in the Apollo missions to the moon. If this were a prop for a science-fiction film, then why mask the man's face? The photographer or origin of the photo remain unknown.

Metallurgy. This S.E.M. image magnified 4000x is of the inner surface shrinkage cavity formed on solidification of a nickel-tin eutectic alloy sphere levitated in microgravity aboard the space shuttle. To capture the fine detail, Polaroid Type 55 black and white instant film was used.

Note the pyramidal structure of this metal alloy used in a Polaroid advertisement.

MOON FACTS

Diameter: 2,160 miles
Mass: 1/81 that of Earth
Volume: 92.4 billion cubic miles (1/49 that of Earth)
Distance from Earth: 240,000 miles
Temperatures of surface rocks: Sun at zenith: approx. 214°F
Night: approx. –250° F
Rotation: Because the moon rotates in the same length of time that it takes to orbit the Earth, we always see the same face of the moon from the Earth.

■ **Viewing the Earth**
Swirls of white clouds could clearly be seen against the blue of the sea and sky, and the brown of the continents.

■ **Footprints**
The footprints left behind by the 12 astronauts who landed there, will be preserved for centuries in the weatherless environment.

■ **Gravity**
Because of the moon's low gravity (one-sixth of the Earth's), the astronauts at first found it difficult to move around. They soon found that it was best to move in a kind of loping gait, rather like a kangaroo.

■ **Rocks**
Moon rocks are different in composition from Earth rocks, implying that the moon almost certainly did not come from the Earth. Samples from the lunar plains are much younger than those from the highlands, which are thought to be part of the moon's original crust.

Source: Pictorial Guide to the Moon, Space & Exploration, Astronomy AP graphic

The moon is thought to be older than the earth, from another solar system, and it is a controvertial point as to whether the moon has only one sixth of the gravity of earths. In actuall 1/6 gravity, it would be easy to move around rather than "loping," and early commentators had projected "slow back-flips on the moon".

NOBODY
GETS
OFF
UNTIL
THE
MESS IS
CLEANED
UP.

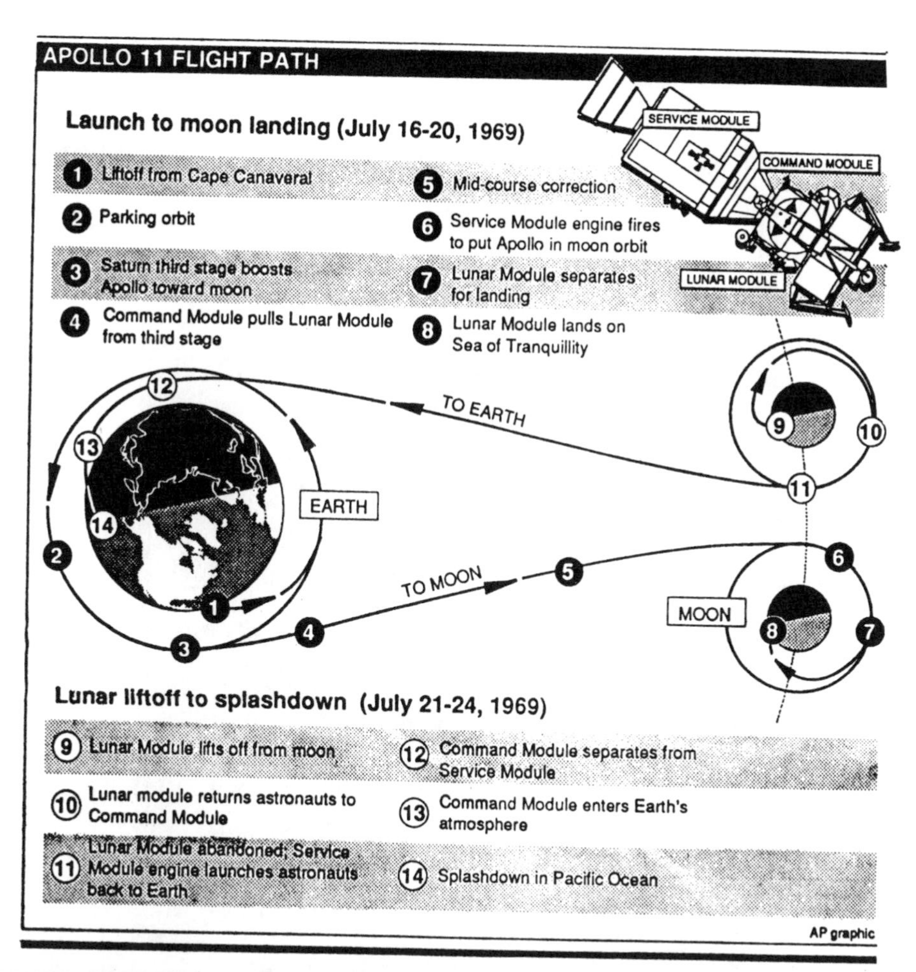

APOLLO 11 FLIGHT PATH
Launch to moon landing (July 16-20, 1969)
1 Liftoff from Cape Canaveral
2 Parking orbit
3 Saturn third stage boosts Apollo toward moon
4 Command Module pulls Lunar Module from third stage
5 Mid-course correction
6 Service Module engine fires to put Apollo in moon orbit
7 Lunar Module separates for landing
8 Lunar Module lands on Sea of Tranquillity
SERVICE MODULE
COMMAND MODULE
LUNAR MODULE
TO EARTH
EARTH
TO MOON
MOON
Lunar liftoff to splashdown (July 21-24, 1969)
9 Lunar Module lifts off from moon
10 Lunar module returns astronauts to Command Module
11 Lunar Module abandoned; Service Module engine launches astronauts back to Earth
12 Command Module separates from Service Module
13 Command Module enters Earth's atmosphere
14 Splashdown in Pacific Ocean
Dangerous approach to Tranquillity Base
At 500 feet, astronauts get first clear view of landing site.
Armstrong takes manual control, flying forward to avoid the rough terrain.
Eagle hovers at 200 feet while Armstrong looks for a safe area to set down with 60 seconds of fuel left.
Eagle touches down, four miles off target, with about 12 seconds of fuel left.
On-board computer was taking Eagle into boulders surrounding crater the size of a football field.
AP graphic

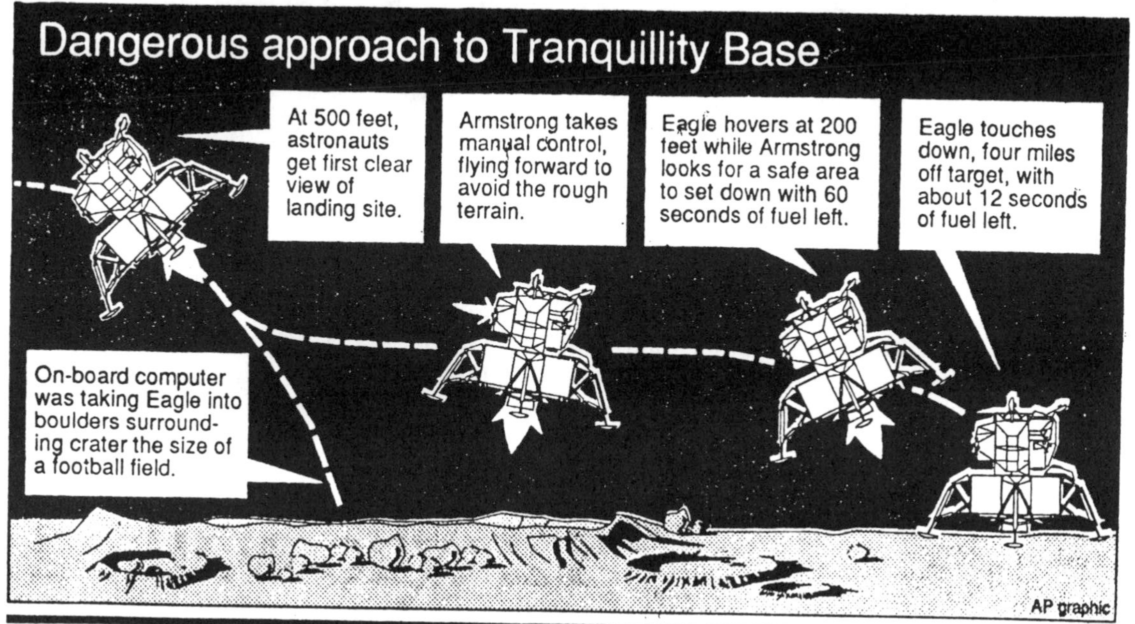

Were Ant-Gravity devices used on the manned Apollo flights to the moon? The Lunar Module when landing on the surface of the moon and taking off acted in a very erratic and strange manner. Official NASA footage of the landings are enough to convince any critical viewer. Note this recent AP graphic drawn for the 20th anniversary of the landing, it shows the lunar module landing on the moon and remanuevering in such a way that is just physically impossible for rocket power.

Searching for Einstein's brain at the Cornell Brain Collection. They know that its around there somewhere...

COURTESY DR. J. ALLEN HYNEK

Laboratory analysis showed considerable chemical changes in this soil sample taken from a UFO landing site near Delphos, Kans., in 1971. The soil no longer absorbs water.

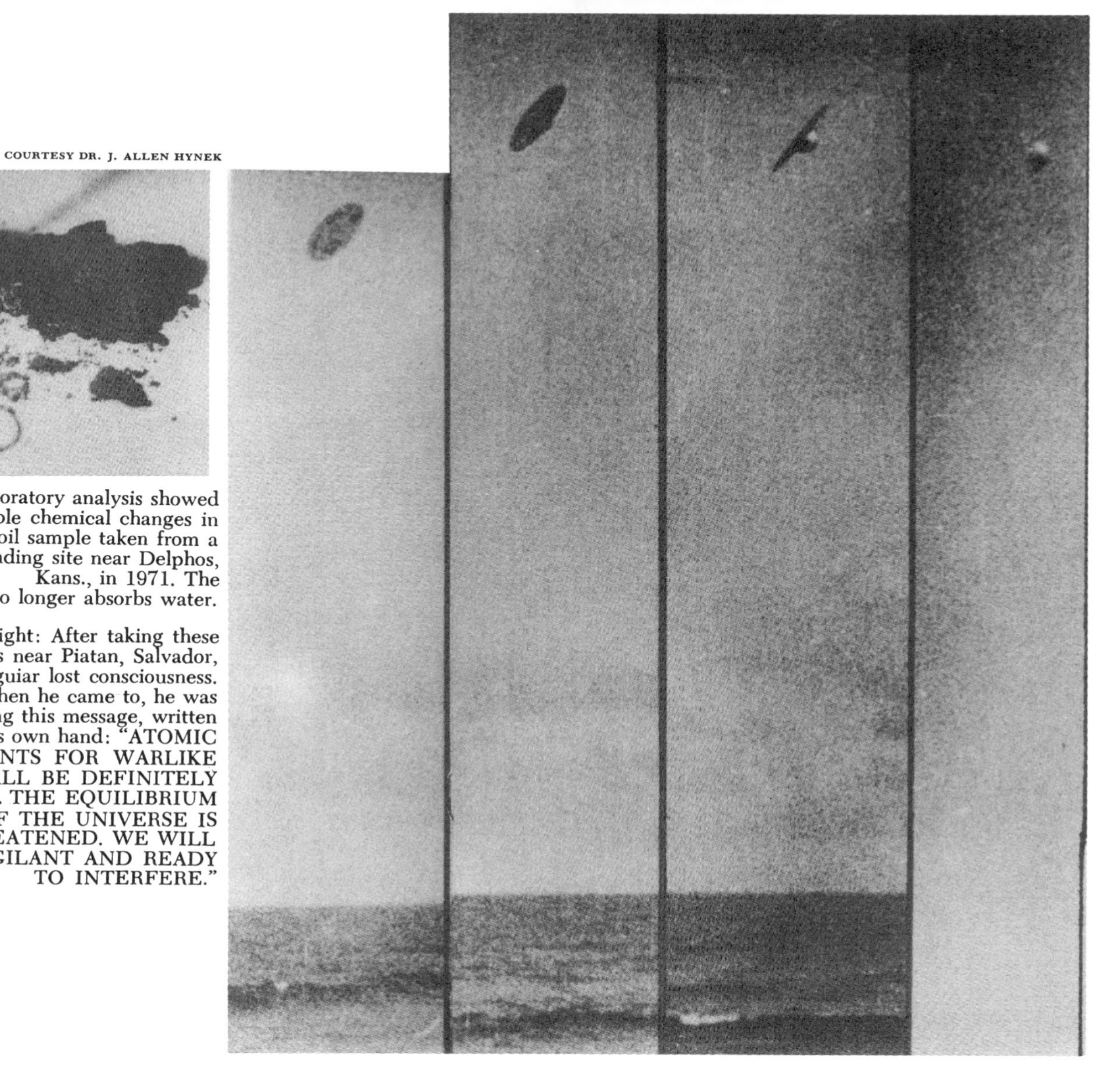

Right: After taking these photographs near Piatan, Salvador, Brazil, Helio Aguiar lost consciousness. When he came to, he was clutching this message, written in his own hand: "ATOMIC EXPERIMENTS FOR WARLIKE PURPOSES SHALL BE DEFINITELY STOPPED . . . THE EQUILIBRIUM OF THE UNIVERSE IS THREATENED. WE WILL REMAIN VIGILANT AND READY TO INTERFERE."

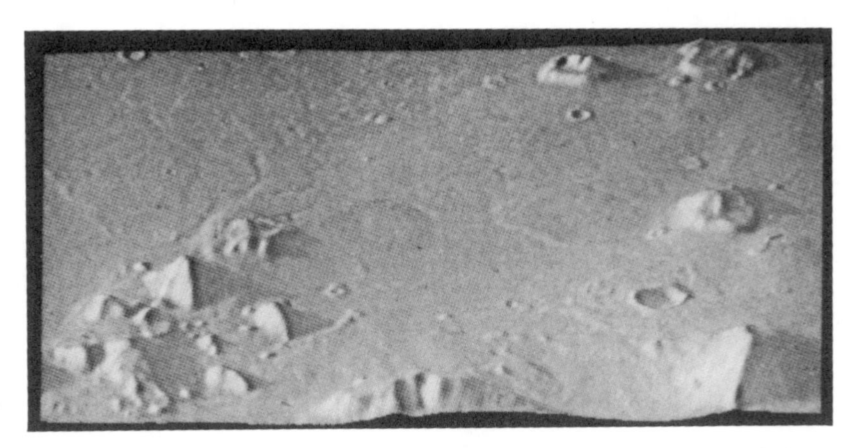

Perspective view of the face on Mars (top right), the "city" (left) and the D & M Pyramid (lower right), looking from the west and well above the Martian surface. Processed by Dr. Mark Carlotto.

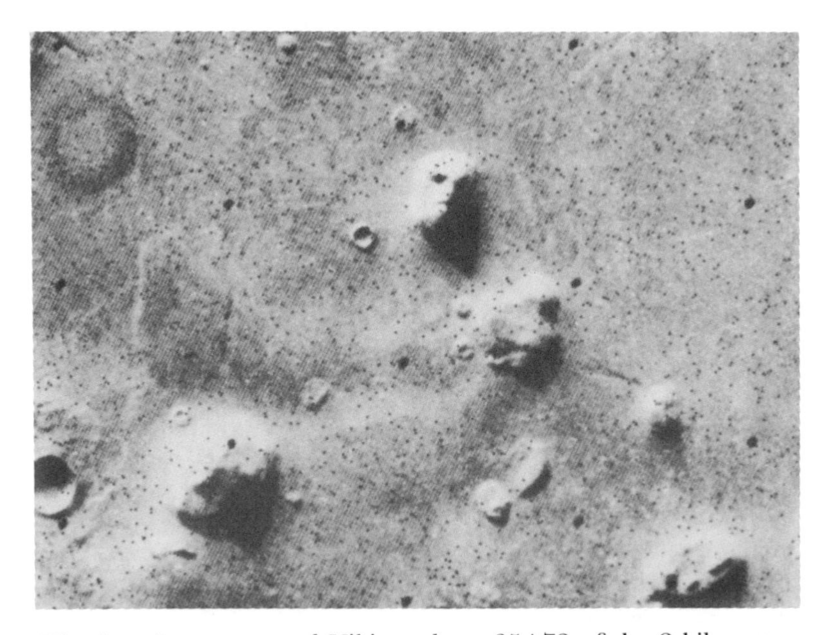

The "raw," unprocessed Viking photo 35A72 of the 2-kilometer wide face on Mars and the surrounding terrain. The circle in the upper left corner is a blemish on the television camera. Black and white specks are noise spikes in the data transmission.

Above, cigar-shaped object photographed over Peru by a farmer in 1952.

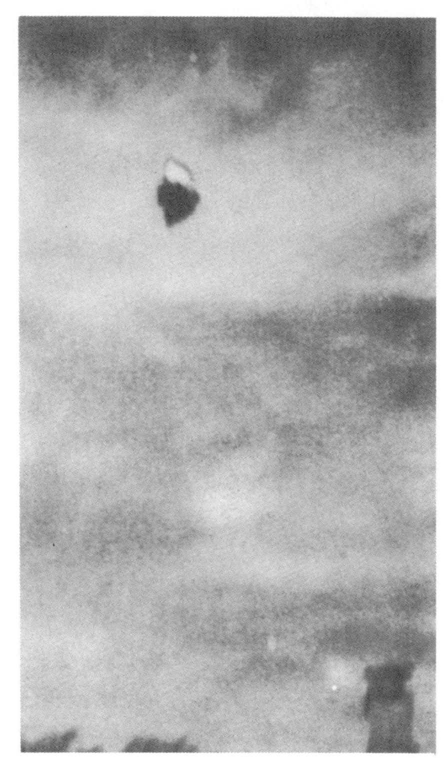

Right, UFO photographed in 1966 by business executive near Melbourne, Australia.

U.S. Airforce photo of a UFO. One wonders if it is not one of their own? UFOs are most commonly seen around Military Bases and Powerstations. Photo courtesy of PROJECT Blue Book.

PROFILE: STELLAR AEROSPACE INDUSTRIES

"When thinking of your personal spacecraft... think *Stellar Aerospace.*"

Gravity control, electromagnetic pulse drives, over-unity generators and the Unified Field have not been put into a technological use for the consumer... until now. The Stellar Research Institute and its manufacturing division, Stellar Aerospace Industries, are putting together the first commercially available anti-gravity spacecraft. For years, talk of privately manufactured flying saucers and anti-gravity powered airships has been existed in the press. Generally, those corporations that worked on such projects were either working in conjunction with the government or were one way or another bought out by government contractors. It seemed as if civilian and non-military projects to develop this technology was just not meant to be.

One organization that has forged ahead in developing peaceful spacecraft for the good of all mankind has been the Stellar Research Institute. The Institute is a not-for-profit organization of scientists, engineers, computer and aviation experts, researchers and business men. Affiliated with the Stellar Research Institute is its for-profit industrial and manufacturing wing, Stellar Aerospace Industries. The prototypes of the spacecraft and airships developed by Stellar Research are then put into production and manufactured at the various facilities of Stellar Aerospace.

A third entity associated with Stellar Research is the Stellar Aerodrome, a one mile square facility to be located in Illinois. The Stellar Aerodrome will be the main airfield and base for the spacecraft and airships manufactured and operated by Stellar Aerospace. Plans for the Aerodrome include a 500 foot high pyramid control tower, gigantic hangars for airships, and a large field for the vertical takeoffs and landings of the craft.

According to Peter Braniff, publicity coordinator for all three groups, the airships will use a "gravitational pulse drive" to move the ship forward and counteract the earth's gravitational pull. An undisclosed "electro-magnetic free energy motor" will power the "gravitational pulse drive." When asked whether any of the devices used are currently patented, his answer was negative.

When asked whether the Stellar Research Institute would seek a patent for any of the devices used in its craft he replied, "Probably not. Patenting these devices is often foolish. The patent office takes a dim view of these devices and were a patent to ever be issued, the military has first option on any invention and can forcibly buy the invention and/or place a gag order on the inventor or corporation that is seeking the patent. We don't want that to happen. We believe that the military and NASA, under the secret direction of the National Security Agency and MJ-12, already has these devices, or a version of them. We are not seeking to sell or patent these devices. We will keep them an industrial secret for awhile and then probably release the plans to the public. This is a technology that we want the world to have. Too many of these sorts of inventions never make it out simply because of greed and the belief by the inventor that he or she will not get hius or her just dues for an invention. As a result, no one wins, except those who wish to continue the suppression of this important technology."

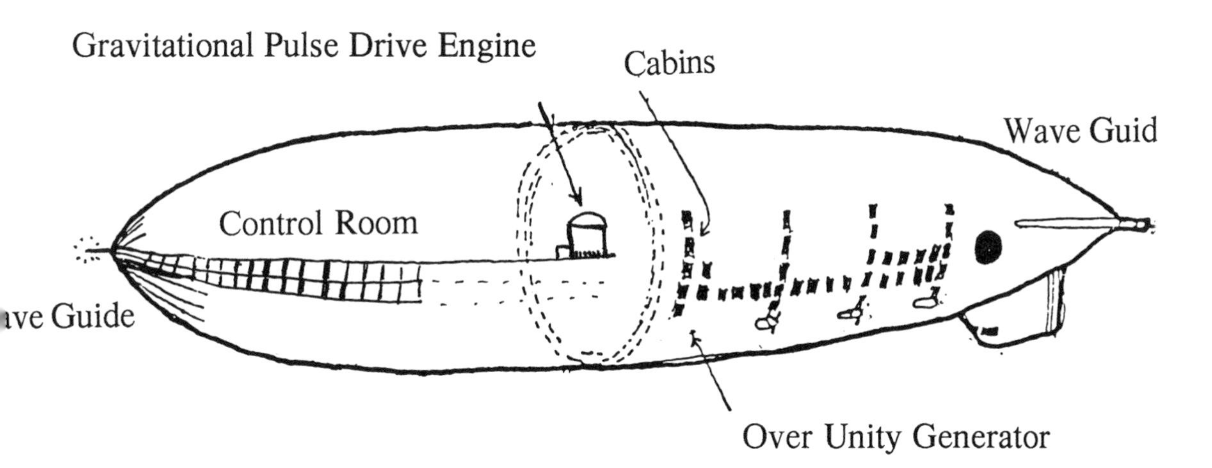

Stellar Aerospace Industries SAI 222 airship.

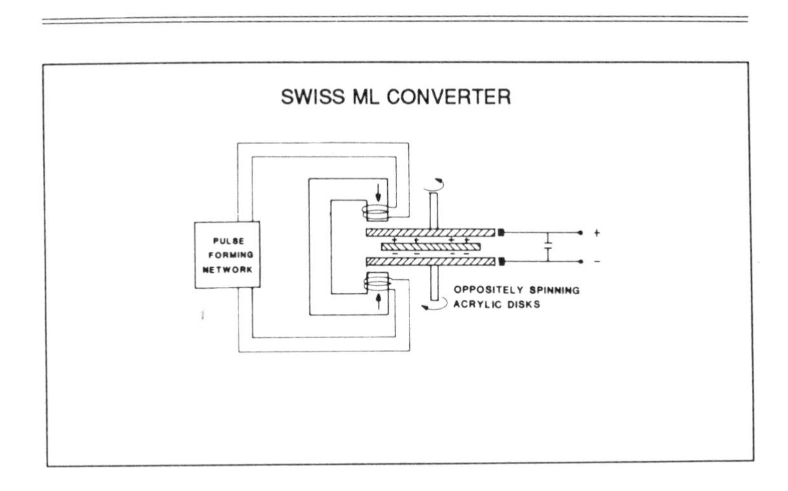

Simplified diagram of a portion of the powersource for the airships.

Prototype airships above the Thar Desert in India, 1985.

When asked more about the planned "Gravitational Pulse Drive" and how it works, Braniff said, "Opposing electromagnetic fields in *pulsed field opposition* produce a direct interaction with orthogonal Zero Point Energy flux. When electromagnetic fields oppose, there is not net field vector, yet the stress on the fabric of space increases. The electrons arrange themselves on the skin of the cage such that all field vectors inside are in perfect opposition. When the charge is increased the interior field vectors increase in magnitude but remain in opposition. Quantum gravity theories show that a high potential or large components of the stress energy tensor can alter the action of the Zero Point Energy

"If abrupt bucking fields were impressed around the nucleus or ion lattice, a direct orthorotation of the Zero Point Energy flux occurs. The leading edge of the pulsed fields pinches the orthogonal flux and builds a gravitational pressure; the sudden release allows the energy, a partial gravitational field, to snap-back into our 3-space. And so the craft moves by so-called Gravitational Pulse. That is all that I can say."

Plans by Stellar Aerospace and Stellar Research are mind-boggling to many. Based out of the high-tech, self-sufficient city of Stelle, Illinois, Stellar Research is planning to link no less than five or more intentional communities around the world by regular airship routing, as well as linking these various communities to L-5 orbit in space.

The city of Stelle was founded in 1971 by a group based in Chicago. Today, the city has a population of about 200 with its own telephone company, reverse osmosis water treatment plant, school, sewage treatment, gasohol plant, hydroponic greenhouse, manufacturing plant, circuit board design company, metal machining industries, injection mold factory and other industries. Despite the amazing claims of Stellar Research, Stelle is a very real and thriving high-tech, alternative city.

Planned Stellar Aerospace facilities in north-central Illinois.

Stellar Aerospace airship at the aerodrome.

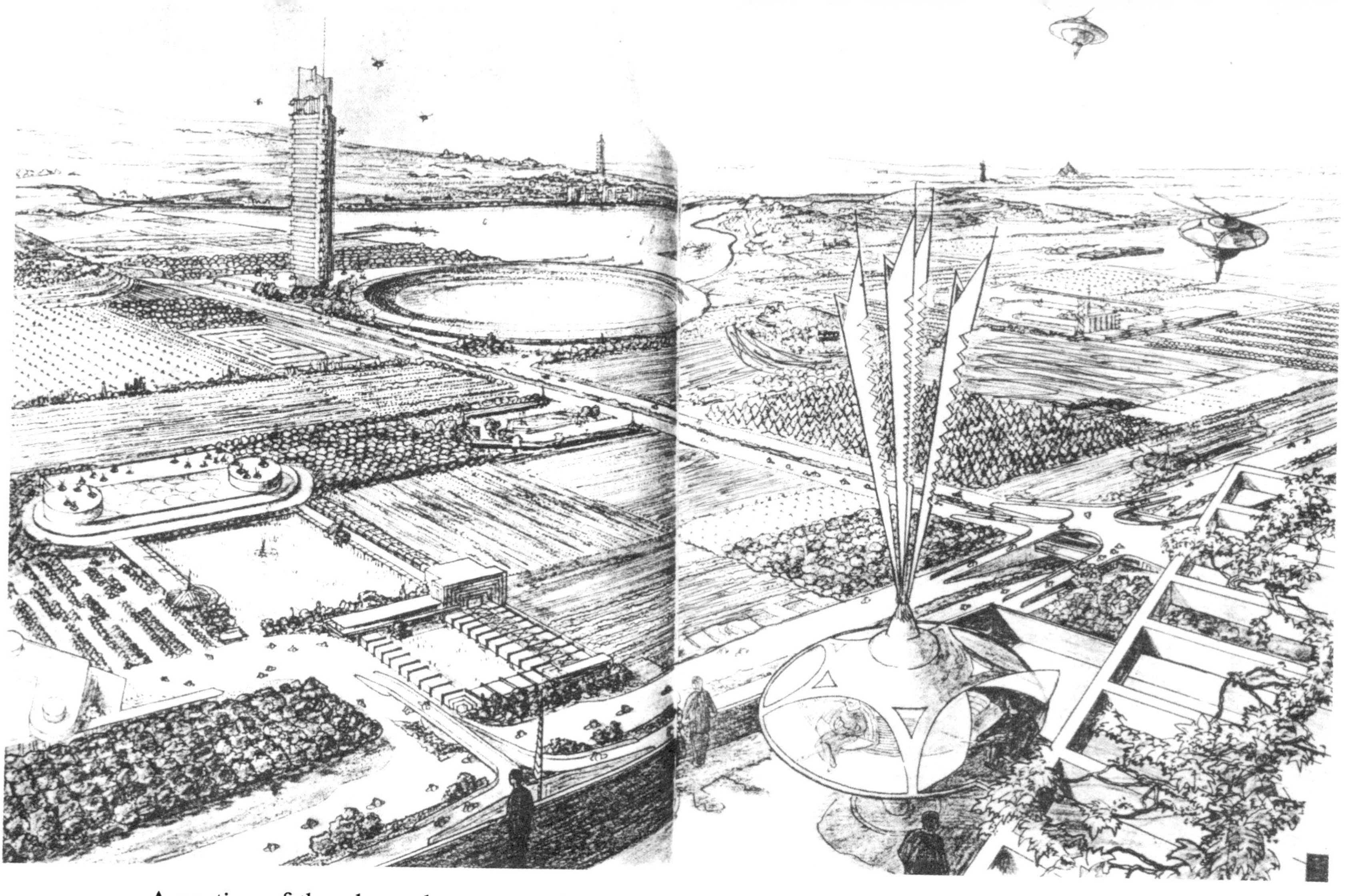

A portion of the planned aerospace facility and space port at Stelle.

Planned city layout for Stelle, Illinois. Less than 10% has been completed.

Stelle, Illinois. This photograph appeared in *US* Magazine, 1977.

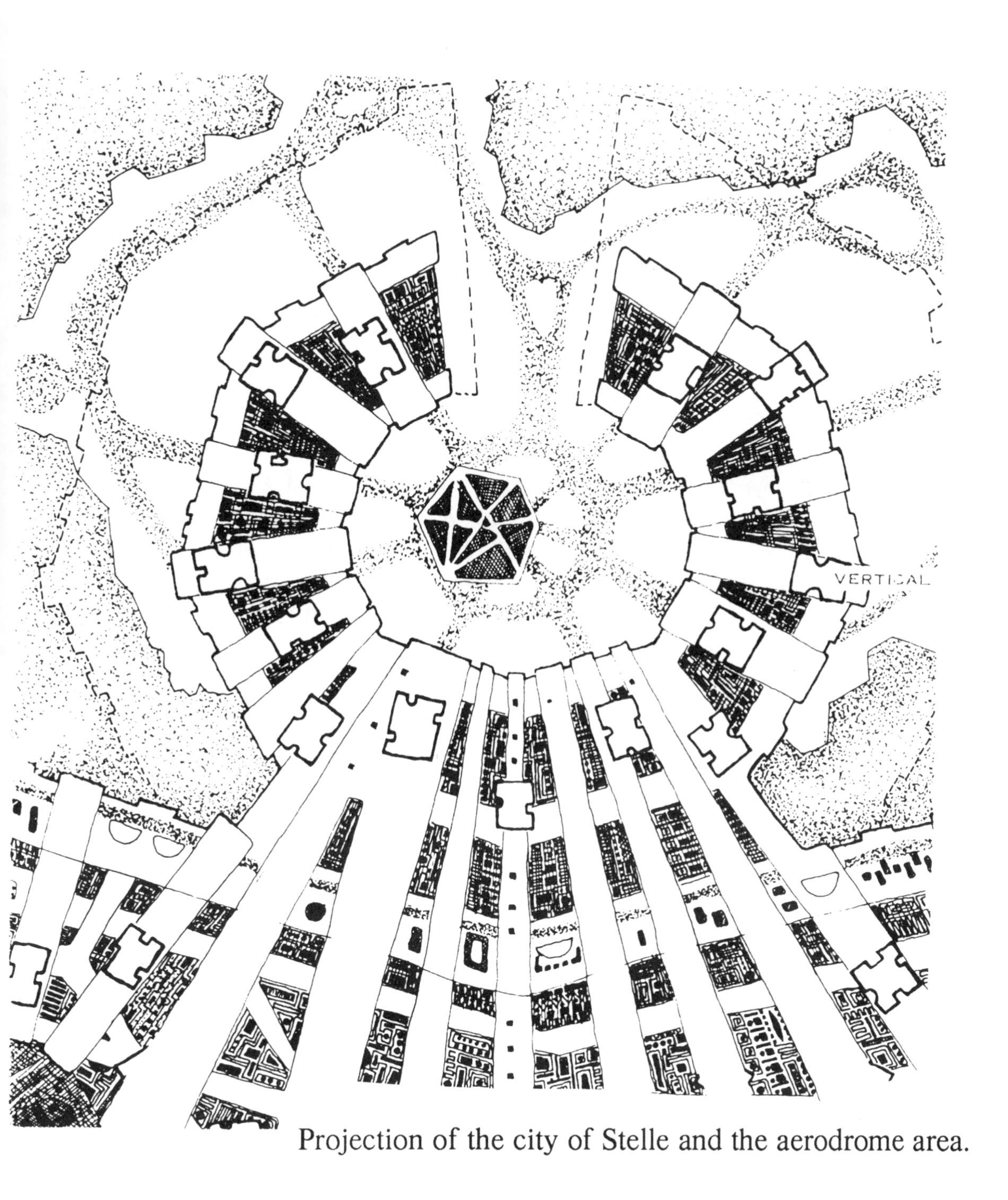

Projection of the city of Stelle and the aerodrome area.

"Most people at Stelle are aware of the airship and anti-gravity plans. Projects of this nature having been going on there for twenty years," said Peter Braniff. "Stelle is not some mythical or secret city of Initiates. It is very real and populated by right-thinking, practical people with normal jobs and children and a desire for a better life. If they have any faults, it is that they are too idealistic."

Braniff says that other alternative communities like Stelle are being built, and they will be linked via the airships manufactured and designed by Stellar Aerospace and based at the Aerodrome in Stelle. "One such community is in Brazil. This city, still in its inception, is called the Cidade de la Paz, which means the City of Peace in Portuguese. They already have designs for the aerodrome there. The city is to be built in the Roncador Mountains of central Brazil, and will be a self-sufficient city similar to Stelle. We plan to help them with the construction and funding of the city, and we will operate an airship run there, probably once a week.

"We plan to link the Cidade de la Paz with the Pyramid Research Center in the Peruvian jungle. Like the Cidade de la Paz, this city is still in its infancy. Another community that we plan to link is Sedona, Arizona. While this is already a thriving town and not a small, self-sufficient community, we believe that it would be the best place to link with near the West Coast. The city of Arcosanti, presently being constructed also in north-central Arizona, could also be used as a stop-over. Because of the nature of this area, we will probably be gathering passengers for our Pacific route. Our airship route would then continue across the Pacific to an undisclosed location in the central Pacific. This location will a be a high-tech city-state located, probably, on land which has yet to rise south of Hawaii. Our research and intelligence informs us that by the mid-90s this area should be available for development by certain special groups.

"Another area in the Pacific that we plan to link up via

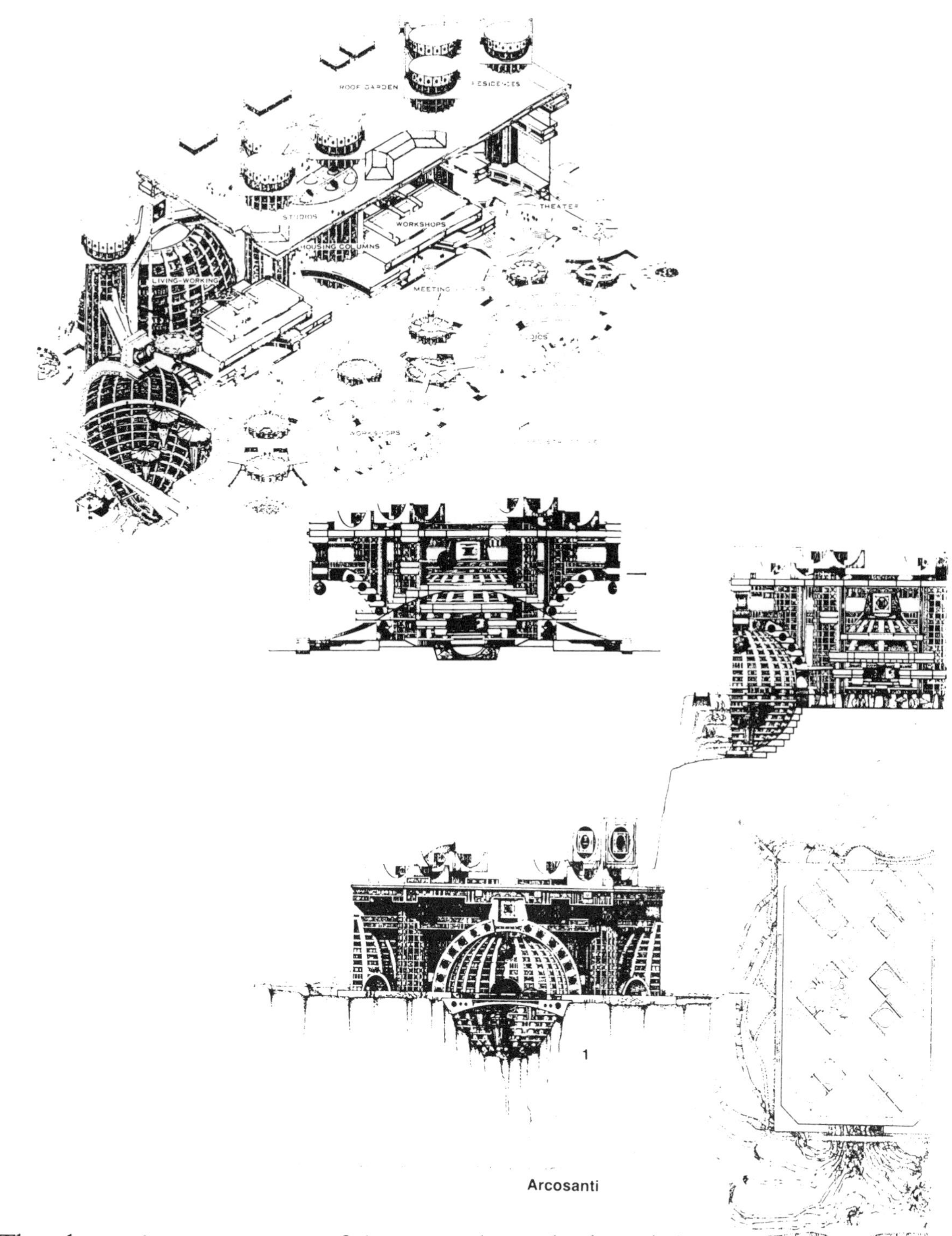

The alternative community of Arcosanti being built in Arizona.

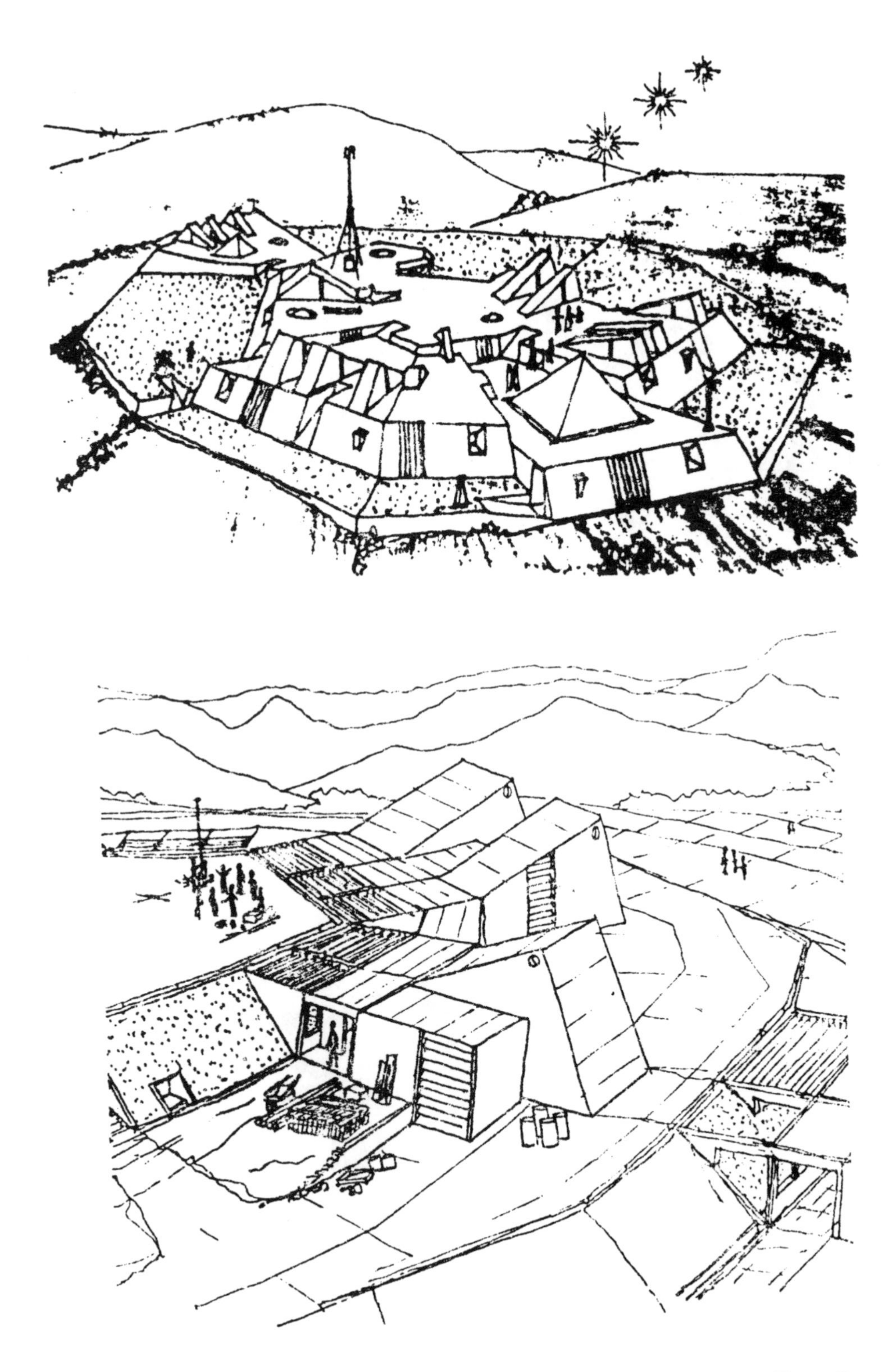

Drawings for the aerodrome at the Cuidad de la Paz in Brazil.

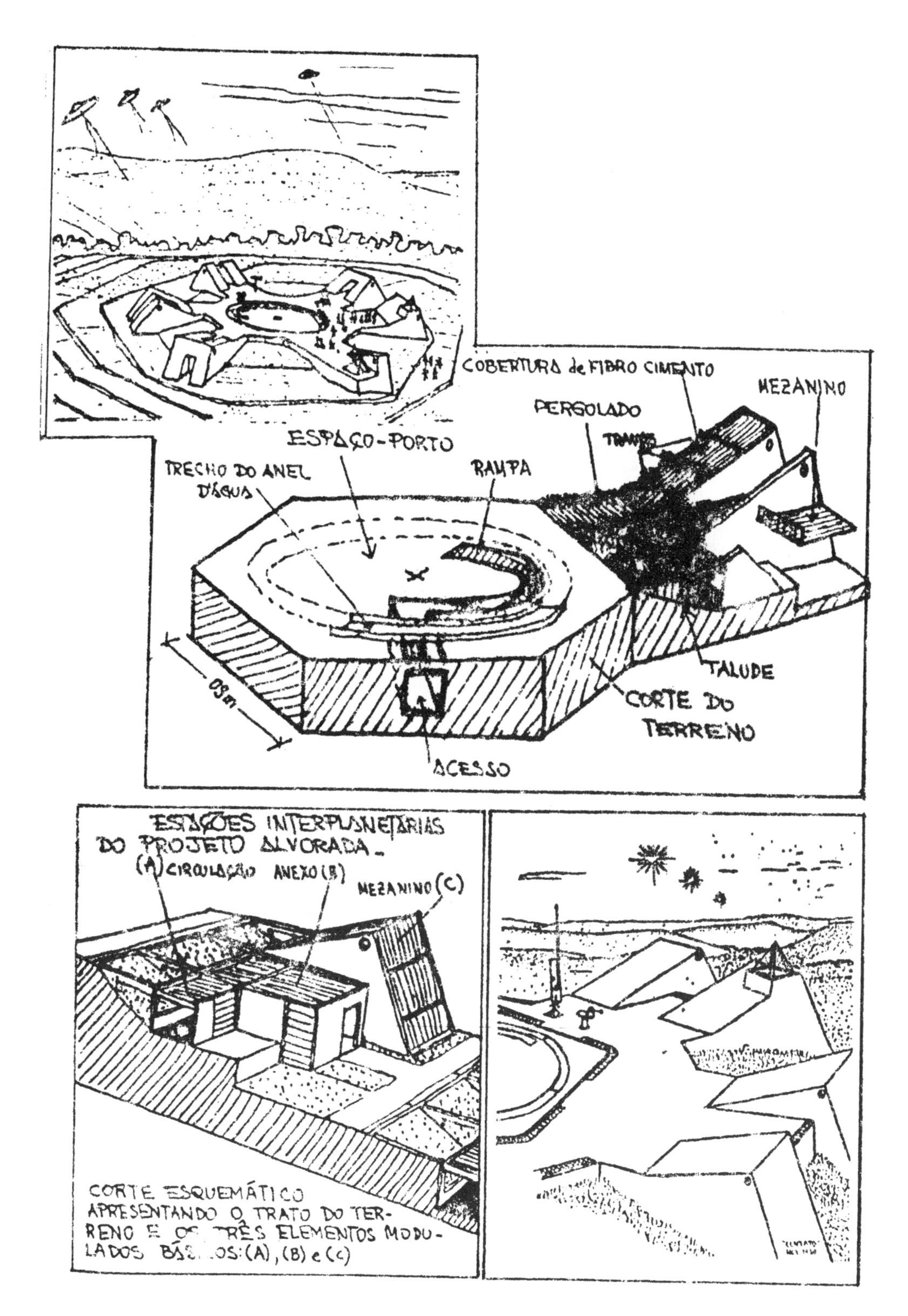

Drawings for the aerodrome at the Cuidad de la Paz in Brazil.

Discoid craft landing at the aerodrome in Brazil.

airships is Easter Island. Using the massive space-shuttle landing area on the island, we will use Easter Island to help link our South American city-bases with the community in the Central Pacific. Further links are to be established with communities in New Zealand and Australia. These communities have not yet been established, as far as we know, but one will probably be in the Blue Mountains of Australia. Because of the likelihood of increased earthquake and volcanic activity during the next fifteen years, we would like most of airship bases to well away from coastlines, and to be a relatively high altitude. For the Pacific Islands that we intend to link, however, this obviously cannot be the case. Airships will be maintained at the mid-Pacific island city for emergency airlift if the island is threatened by a tsunami or unusually large tidal wave."

Why the stress on earth changes, tsunamis and airlifts? Braniff replies, "It seems likely that we are about to enter a period of intense geological activity, far more active and destructive than most geologists would like to believe. This increased activity of geological change may even culminate in a so-called 'pole-shift' at the turn of the century. If that is the case, we want to be ready for it. During poleshifts, nearly every volcano in the world blows, oceans spill out of their basins and wash across continents.

"Tectonic plates rise and fall and hurricane force winds descend onto the earth. It's a tough time for everyone. Logically, an L-5 orbit around the earth as in Gerard O'Neil's concepts would be the best place to hang out during this time. Therefore, we intend to be able to offer customers that the chance of going into space during that period. Essentially, we hope to run one-to two-week tours, holidays if you will, into space."

Does that mean that the Stellar Research Institute plans to build a hotel in space?

"Our airships are hotels. They are mobile hotels. Each airship can accommodate about 500 people and be in space indefinitely. Since we do not use conventional fuel, we have an unlimited range and time per flight.

Said Ethel P. Zufelt, former secretary of **SRI** in a recent interview, "I remember once when Winston [referring to the enigmatic scientist Winston Whitaker, who has done some designing for Stellar Research and Stellar Aerospace] was working on perfecting his gravitational pulse drive and had some accident. He came running out of his lab and said that there was a sudden danger of EMP and not to touch anything metallic or we might be electrocuted. I guess something must have gone wrong."

Said Mrs. Zufelt, "I never knew that much about him, but I remember one time when he mentioned how he flew his first flying disc from his lab in central India to somewhere in Tibet ...and I think the border was supposedly closed at the time!"

Information on Whitaker is not forthcoming from the usually gregarious bunch at Stellar Aerospace/ Institute. "Its not that we mind talking about Winston," said Peter Braniff, "Its just that we don't know what to say. We don't even know where he is."

Wherever Whitaker's whereabouts, his *Gravitational Pulse Drive* is a large part of Stellar Aerospace's plan for their trips to L-5 orbit. The Gravitational Pulse Drive is the main power unit for the space craft, and currently has not been patented. "We see no need to patent our innovative *Gravitation Pulse Drive* (GPD), for various security reasons. Patenting is not necessarily the best way to go. We prefer the other method of what is known as an industrial secret. However, it is not our objective to sit on this technology; we are for the introduction of this technology all over the world, especially to the layman and man-in-the-street."

In a rare interview in 1988 H. Gordon Florim, senior engineer at Stellar Research said, "We intend to build four space-ports around the world and one space hotel-space dock. The first space-port will be built just south of Stelle,

Linking of the alternative communities via airship. 1. Stelle, 2. Sedona, 3. Central Pacific, 4. Pyramid Research Center, 5. Cuidad de la Paz, 6. Easter Island, 7. Blue Mountains of Australia.

Pyramid Research Center in the central Peruvian jungles.

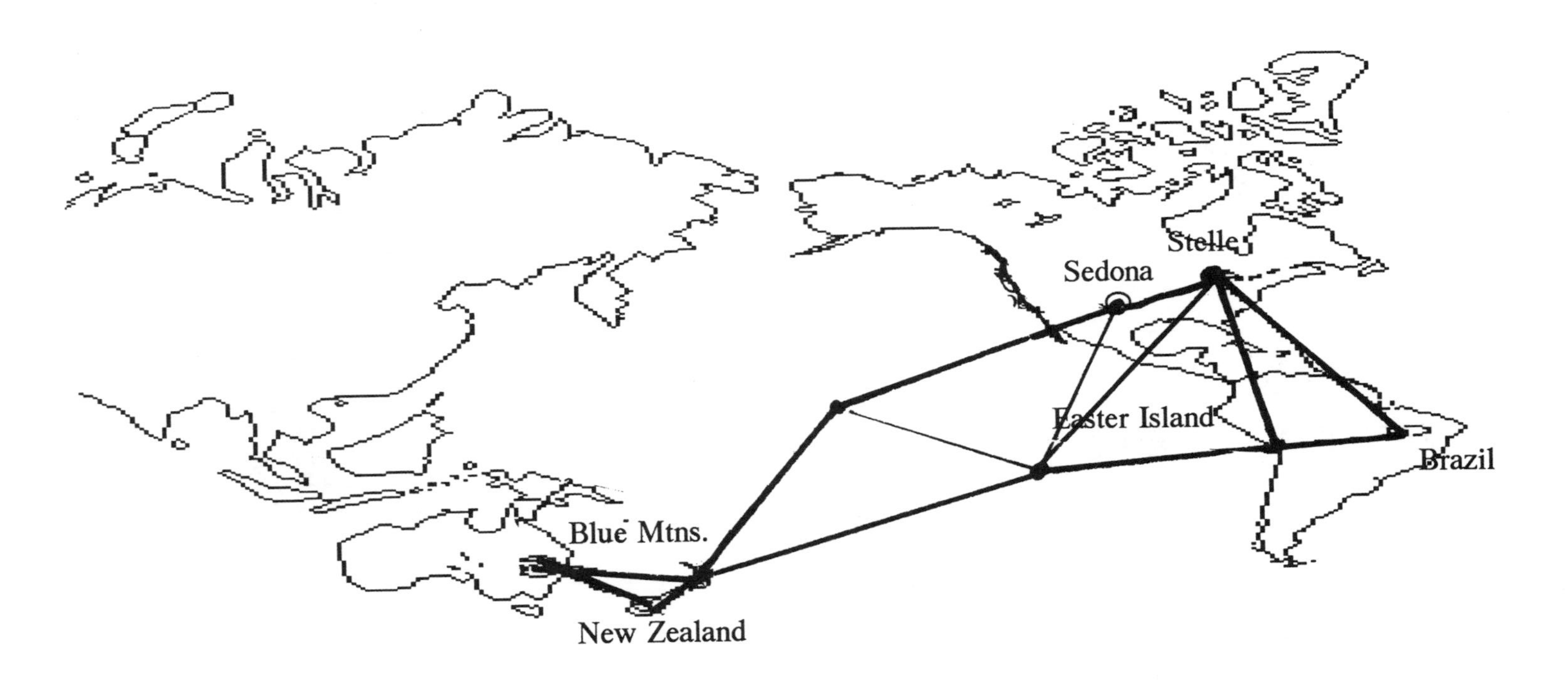

The projected airship routes between Stelle, South America, & the Pacific.

Illinois, tentatively named the *Stellar Aerospace Stardock.*

"The second space-port will be built in the Roncador Mountains of central Brazil. The name of the city-space-port will be *"Cidade de la Paz."* The third space-port will be built in east-central Australia.

"We will construct and operate large, electro-magnetically-driven ships. Certain motors and technology will be contracted out. Though we are currently planing to buy old airliners, remove the wings and install our own anti-gravity motors, we have contracted to an Australian company who will manufacture the aircraft frames.

"Our current plans are to construct one space hotel-space dock in L5 orbit. This may be done in conjunction with Hilton Hotels and their multi-million dollar contract now being offered. Each space-port will have a self-sufficient community built around it.

"The fourth city will be built on (name deleted) presently 350 ft. below the Pacific. This newly risen island will be occupied and declared an independent nation by staff of the Stellar Research Institute/Stellar Aerospace Corporation.

"We will network all of the cities, and others, by our airships, products and services. Immigration and emigration from all over the world to these four cities will occur. Population of the city at (name deleted) will come from selected communities in North America, Brazil, Australia, Europe, India, China, Africa and elsewhere.

"On April 29, 2000, we will conduct a massive airlift into space for two weeks at each of the four cities. It promises to be an exciting event."

Stellar Aerospace plans to manufacture two varieties of Airships, the largest being the cigar-shaped SAI-222. A smaller discoid, the SAI-112 will be a shuttle and personal transportation vehicle.

THE SAI-222

•ELONGATED CYLINDRICAL SHAPE
•5-PERSON CONTROL ROOM
•LARGE STORAGE AND LIFT CAPACITY
•FIBERGLASS AND ALUMINUM CONSTRUCTION.
•NON-MAGNETIC HULL
•RESONATE SOLID STATE POWER SOURCE
•FIVE-CRYSTAL HARMONIC IMPULSE DRIVES
•12 GRAVITATIONAL-WAVE GUIDES FOR STEERING
•INTERPLANETARY CAPABILITY (HYPERSPACE DRIVE EXTRA)
•CAPACITY 8,000 PERSONS
•VERTICAL TAKE-OFF AND LANDING
•BASIC MODEL (STRIPPED DOWN) $5,384,850.00 GOLD STANDARD DOLLARS.
•TAX AND DELIVERY NOT INCLUDED

THE SAI-112

•DISCOID SHAPE WITH DOME
•FIBERGLASS AND ALUMINUM CONSTRUCTION
•NON-MAGNETIC FIBERGLASS HULL
•RESONATE SOLID STATE POWER SOURCE
•SINGLE-CRYSTAL HARMONIC IMPULSE DRIVE
•THREE FIELD-STABILIZER DOMES WITH A CENTR WAVE-GUIDE FOR EASY HANDLING
•INTERPLANETARY CAPABILITY
•CAPACITY: 5 PERSONS
•VERTICAL TAKE-OFF AND LANDING
•BASIC MODEL (STRIPPED DOWN) $1,000,850.00 GOI STANDARD DOLLARS.
•TAX AND DELIVERY NOT INCLUDED

For more information write to:

Stellar Research Institute
Box 74
Kempton, Illinois 60946 USA

The Pyramid Research Center
Box 5271
Ft. Lauderdale, Florida 33310 USA

The Biodome Project
Windstar Foundation
Box 286
Snowmass, Colorado 81654 USA

Arcosanti
Cosanti Foundation
6433 Doubletree Rd.
Scottsdale, Arizona 85253 USA

For information on the community of **Stelle**, write to:
Stelle Chamber of Commerce, Stelle, Illinois, 60919.

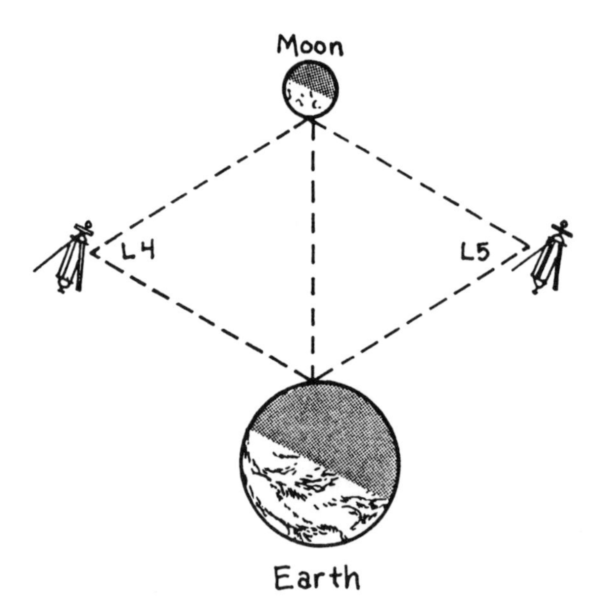

Location of the Lagrange Points L4 and L5. Each is on the orbit of the Moon and is the third point of an equilateral triangle, the Earth and Moon being the other two points. Space communities could be located on stable orbits about either L4 or L5.

Stellar Aerospace facilities planned for L5 orbit.

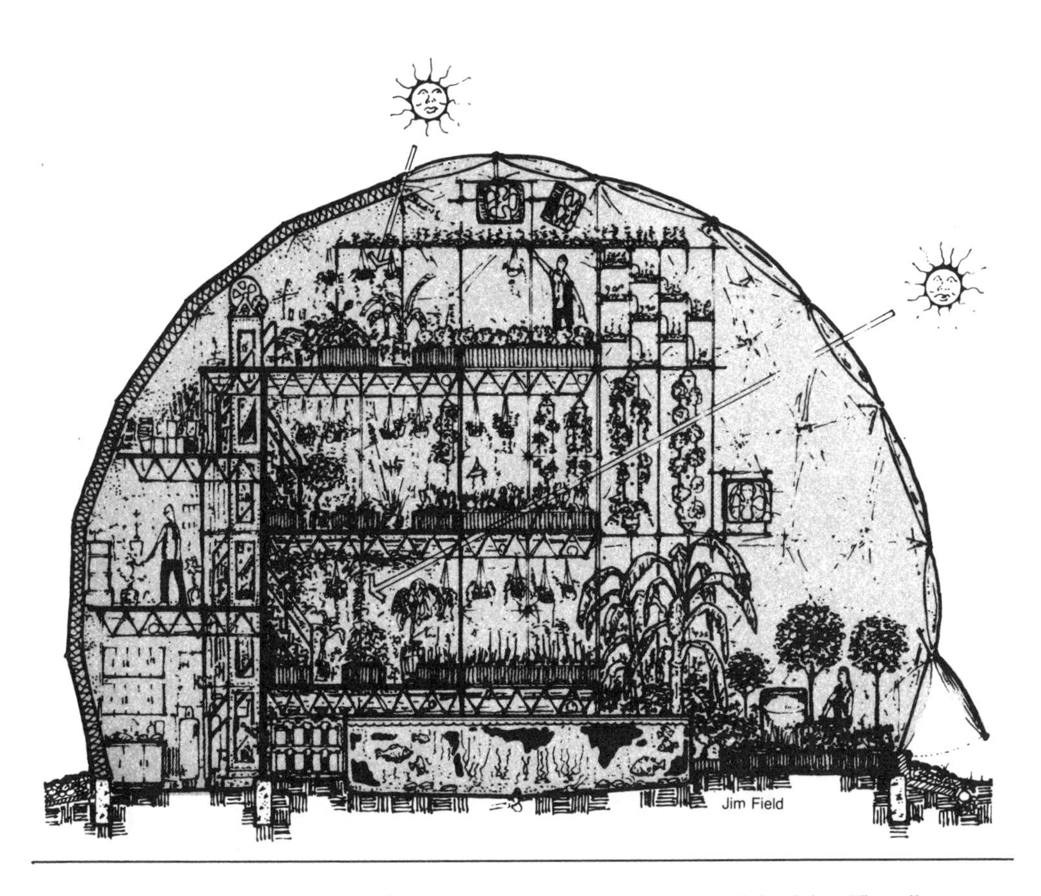

THE BIODOME STRUCTURE

The Biodome design utilizes the concept of the "basket weave" (deresonated tensegrity) dome developed by Buckminster Fuller. This type of dome requires less structural material than a conventional geodesic dome. Windstar developed the innovative connecting hardware which makes it possible to use standard pipe as the main structural component. An inflated plastic double layer skin is attached to the structure creating a durable and effective greenhouse glazing. Dome structures enclose the most volume with the least amount of surface area. By filling the volume of space with planting systems rather than just utilizing the ground floor area, a very high density of food production can be achieved.

The double layer inflated "pillows" are made of new, long lasting plastic materials. The outer skin of the pillow is made of a polyester film coated for high light transmission. The inside layer of the pillow is made of a polyester film with a metallic coating (AgriFilm 88, Southwall Corporation). The inflated pillows allow the sun's energy in while reflecting heat back into the dome. These plastic materials optimize performance factors such as insulating value, light transmission and durability. The pillows have an expected lifetime of 12 years.

The skin of the Biodome can be altered to operate in a wide range of climates. The structure itself will withstand high winds and extreme snowloads. The Biodome can be assembled with hand tools. Each of its structural members weigh less than 20 lbs. and can be positioned by hand. The Biodome has great potential as an efficient, cost effective greenhouse kit structure.

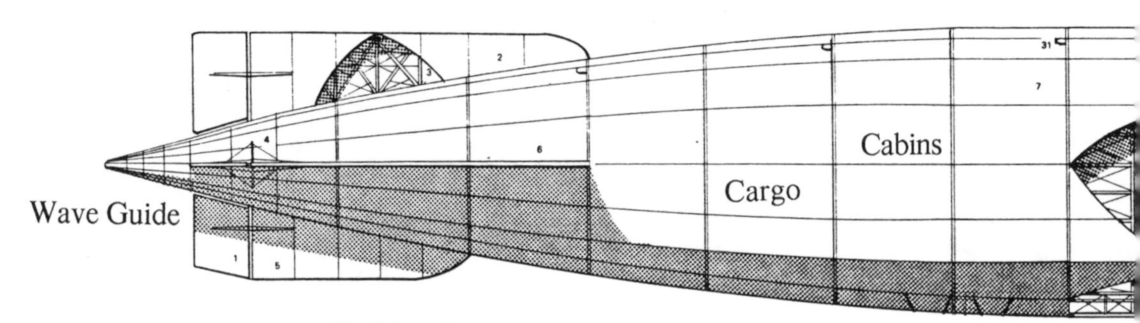

THE SAI-222

- ELONGATED CYLINDRICAL SHAPE
- 5 PERSON CONTROL ROOM
- LARGE STORAGE AND LIFT CAPACITY
- FIBERGLASS AND ALUMINUM CONSTRUCTION.
- MAGNETIC CHROMIUM HULL
- RESONATE SOLID STATE POWER SOURCE
- FIVE CRYSTAL HARMONIC IMPULSE DRIVES

Gravitational Pulse Drive Engines

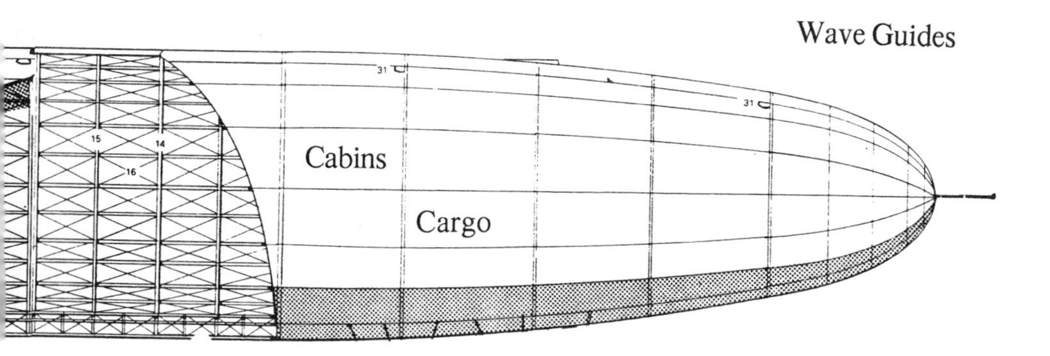

Control Room

LIGHT HARMONIC WARP DRIVE OPTIONAL
***TAX & DELIVERY NOT INCLUDED**

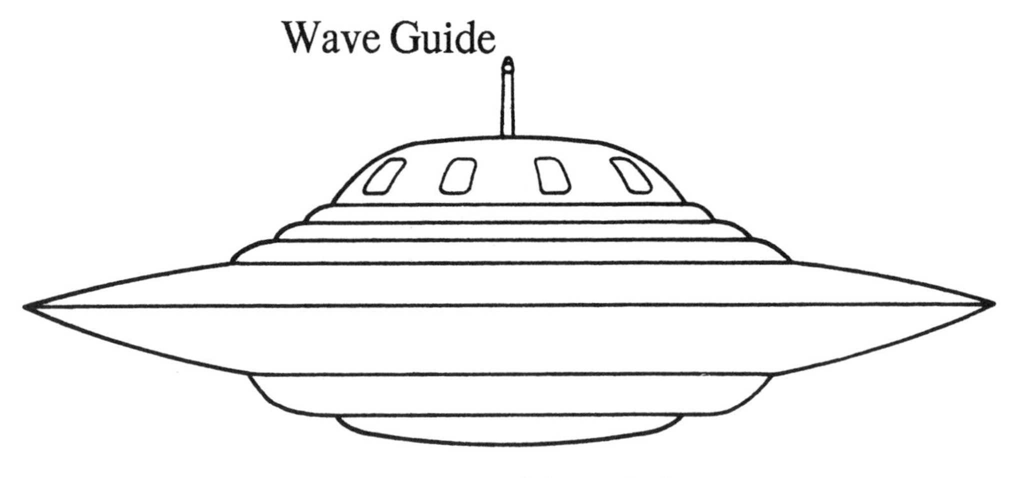

THE SAI-112

- DISCOID SHAPE WITH DOME
- FIBERGLASS AND ALUMINUM CONSTRUCTION
- MAGNETIC CHROMIUM HULL
- RESONATE SOLID STATE POWER SOURCE
- SINGLE CRYSTAL HARMONIC IMPULSE DRIVE
- THREE FIELD STABILIZER DOMES WITH A CENTRAL WAVE GUIDE FOR EASY HANDLING
- INTERPLANETARY CAPABILITY
- CAPACITY 5 PERSONS
- VERTICLE TAKE-OFF AND LANDING
- BASIC MODEL (STRIPPED DOWN) $1,000,850.00 GOLD STANDARD DOLLARS.
- TAX AND DELIVERY NOT INCLUDED

APRIL 23, 2000
UPI Dispatch
AIRSHIPS PREPARE FOR MASSIVE LIFT-OFF!

April 23, 2000. Here at the sprawling Stellar Aerospace Facility several hundred are busy taking off for L-5 orbit for the next several weeks. Professor Winston Whitaker director of the airlift and former chairman of the Stellar Research Institute said that "this isn't just another one of our normal trips into space, we're using all of our airships for this operation. It's our biggest expedition into space that we ever done."

The airships, long and cylindrical like a zeppellin, are powered by the Institutes revolutionary motor, manufactured under license from the inventor. Stellar Aerospace Industries is currently housed on the first and second floors of the massive concrete pyramid at the manufacturing facilities located at Stelle, Illinois.

Said Whitaker, "Our normal cargo service across the Pacific to Australia, and other flights have been suspended until June 1. Our L-5 orbit space camps are our total focus for the next few weeks.

Advanced SAI-222 model for interstellar flight.

THE AGE, Tuesday 26 November 1985

A dream takes to the skies

Blake's baby airship beats Bond's at a fraction of the cost

By GEOFF MASLEN

IT hovers like a ghostly blimp in the gloom of a Kensington warehouse: Melbourne's first hand-built airship.

It has taken Bruce Blake and a dozen volunteers 18 months and hundreds of hours of unpaid labor to build. Today it will be unveiled to the public.

For Mr Blake, it is the culmination of 10 years of trying to get his airship ideas off the ground. Finally he had to resort to a 12-metre, one-third scale model, but his "light utility airship" is now a reality.

The Blake team expect to have their $7500 airship flying over Melbourne before Alan Bond's $6 million version is launched in Sydney next year.

Mr Blake is one of only four airship designers in the world — "that's a pretty exclusive club". He has come up with a novel means to demonstrate the airworthiness of his designs, at minimal cost and with a chance to generate income.

With a band of willing helpers, he spent 18 months building the scaled-down version of the airship in a huge warehouse in Kensington.

The model is intended to be remotely controlled and is expected to be used at outdoor displays such as air shows and sporting events, either as an aerial advertising billboard, or carrying a radio-operated television camera.

Bruce Blake became fascinated with the possibilities of airships after graduating as an aeronautical engineer 10 years ago.

He contributed to the design and construction of Australia's first airship in 1976, the non-rigid Ardath, a 25 metre-long vessel that was later sold to America.

Since then he has worked with the Department of Aviation, the Government Aircraft Factory and the Australian Aircraft Consortium, all the time spending his spare hours working on airship designs.

He calculated the prototype development costs of building a full-size airship at $400,000. He opted instead for a scaled-down version.

Mr Blake approached the Western Australia-based Bond Corporation last year for support but was told that Bond "had a total commitment to the British Airship Industries Ltd" — the firm in which Bond had bought an 86 per cent stake.

"Although I was sorry that Bond was not able to support an Australian initiative, I believe he has shown great foresight in buying Airship Industries," Mr Blake says.

He estimates there is a market for 100 airships in Australia and as many as 1000 in South-East Asia.

Bruce Blake hopes that by showing that his prototype will work, and can generate income, potential investors will help put his plans in the air.

Bruce Blake, with his airship, in a Kensington warehouse.

UNIFIED FIELD FUNNIES

Larry Todd

the sweet smell of science

BENOIT

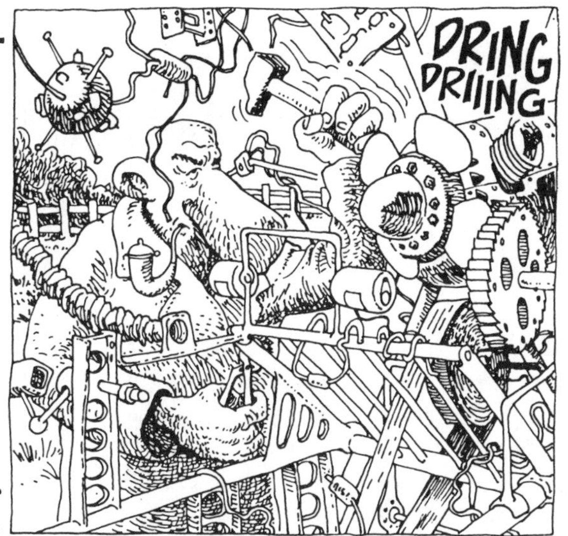

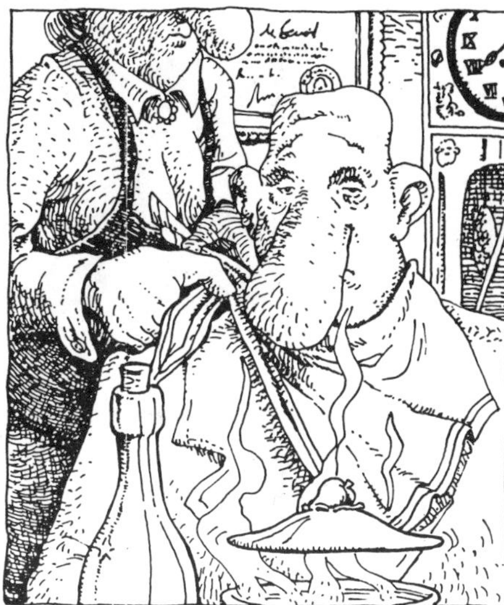

CAREFUL! IT'S HOT!

STOP THAT!

HENRY! I **INSIST** YOU DON'T DO THAT AT THE TABLE!

HENRY!

HENRY!
DON'T!

VRRUUM

VVRRUUMM

WHAT A RELIEF!

FOR TWO DAYS I'VE BEEN HANGING AROUND IN THE PIT WITH A DRIVE SHAFT ON THE BLINK! OL' BUDDY...

EVER SINCE THEY MADE US THAT NEW SET OF POINTS, YOU CAN'T TELL WHAT'S UP WITH THIS ROAD!

AND MY WIFE IS STILL WAITING FOR ME AT THE CLINIC, POOR THING. SHE'S GOT A BUN IN THE OVEN.

YOUR CRATE SURE IS A HELL OF A MACHINE! CYL-INDERS TO THE CAM SHAFT AND ALL THE REST! HO! HO!... COME ON, YOU CAN'T HIDE ANYTHING FROM ME. I'VE GOT AN EXPERT'S EYE.

HEY! IT'S A PLANE, TOO!
FASTEN YOUR SAFETY BELTS!
WHA?!!!

WAI...
LOUISE!
HOSPITAL
H
6

THAT...THAT'S MY WIFE! I'M A FATHER! STO... STOP!

STOP THIS FUCKIN' THING!
LOUISE! WAIT FOR ME!

WELL, IF YOU DON'T WANT TO, I'LL TAKE CARE OF IT MYSELF WITH MY SKELETON KEY!

ONE GOOD TURN OF THE SKELETON KEY AND THAT'S IT!

AND SO: LOUIS ZARZO EXPLORES FUTURE AGES...

WHILE HENRY SCHLITZ-SMITH AND HIS WIFE TAKE IN THE POOR ORPHAN...

JOHN BLAP SINKS A HOLE.

GOOD NEWS IN BOB'S NEWSPAPER:

Calvin and Hobbes

by WATTERSON

GRAVITY IS ARBITRARY!

CALVIN WAKES UP ONE DAY TO FIND HE IS IMMUNE TO THE FORCE OF GRAVITY.

HE HANGS ON TO THE GROUND FOR DEAR LIFE, BUT HIS GRIP IS WEAKENING!

HE CAN'T HOLD ON! HE... HE **LETS GO**!

HIGHER AND HIGHER, AS UPWARD HE FALLS!

ONLY BY GRABBING THE TAIL FIN OF A PASSING JET DOES CALVIN SAVE HIMSELF FROM BEING HURLED OUT INTO SPACE!
WATTERSON

NO, NO, LET HIM FINISH. THIS IS VERY INTERESTING. SO AFTER YOU LANDED IN PHOENIX, WHAT HAPPENED?
WELL, I DON'T CARE. I'M NOT SEWING VELCRO ON THE OUTSIDE OF ALL HIS CLOTHES.
WELL, ABOUT THEN MY GRAVITY CAME BACK, SO I...

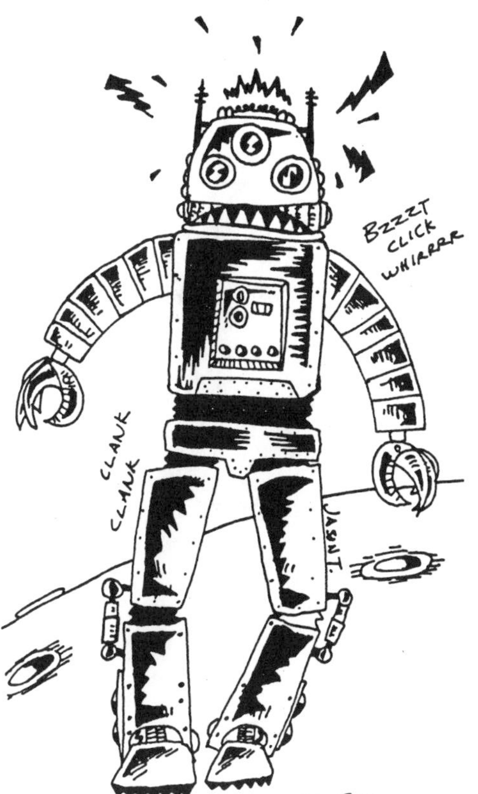

UNEMPLOYED PULP SCI-FI ROBOT IN SEARCH OF STORY IN CYBER-PUNK WASTELAND

DOESN'T MEAN THEY'RE SMART!!!

Channeling, like anything else, requires that the consumer be discriminating and careful. THERE ARE IDIOT ENTITIES, spooks from the other side who are tired of listening to each other. They're desperate to gab, and will do so whether or not their information is valuable or even true!

To protect yourself from flawed or worthless channeled advice, there is TCTS (Trance Channel Testing Systems) Ultima II, an easy-to-administer cosmic aptitude test that enables you to determine whether the entity you consult is a brilliant mind from afar or just another astral couch potato.

Any self-respecting enlightened entity will gladly submit to this simple, verbal test, which takes only minutes. Protect yourself against mumbling fools from the land of the Big Sleep. Remember—JUST BECAUSE THEY'RE DEAD, DOESN'T MEAN THEY'RE SMART!

TCTS Ultima II is sold at fine crystal shops and occult bookstores everywhere.

ULTIMA II

reprinted from *A Search of the New Age* by Chris Kilham (1988: Destiny Books, a Division of ITI; (802) 767-3174)

GEE I'D LIKE TO DO SOMETHING REALLY SIGNIFICANT WITH MY LIFE .... MAYBE BECOME A GREAT WORLD LEADER AND SAVE THE EARTH FROM NUCLEAR DEVASTATION, FEED ALL THE HUNGRY, OR MAYBE EVEN TRAVEL TO FAR AWAY PLANETS .... BUT WHO AM I KIDDING? I'M JUST A TINY BLACK FLY SITTING ON SOME BALD GUY'S HEAD.

これでなに始める気だ？
爆弾でも作ろうてのか？
危ないこといわないの！

It looks mysterious to the women too, but be assured that the man in this Goma Shobo comic book has no intention of building a bomb. Making a high-temperature superconductor in his hotel suite will keep him busy until dawn.

SEZ THE SOVIETS HAVE AGREED TO TAKE JOHN DENVER INTO SPACE NEXT YEAR.
JOHN DENVER? IN SPACE?
10-18

THAT'S WONDERFUL! THAT'S NEWS WE CAN ALL BE JOYFUL IN HEARING! THAT'S--

YOU DON'T S'POSE THEY'RE GONNA BRING HIM BACK DO YOU?

It's science

With Dr. Radium

THAT NASTY, HORRIBLE THING COMING TOWARDS US! WHAT IS IT?!

IT'S SCIENCE!

ANTI-GRAV

THE FREE-ENERGY DEVICE HANDBO

A Compilation of Patents and Reports

by David Hatcher Childress

A large-format compilation of various patents, papers, descriptions and dia vices and systems. *The Free-Energy Device Handbook* is a visual tool for ex magnetic motors and other "over-unity" devices. With chapters on the Adams Mo cold fusion, superconductors, "N" machines, space-energy generators, Nikola Tesl the latest in free-energy devices. Packed with photos, technical diagrams, patents and this book belongs on every science shelf. With energy and profit being a major politic various wars, free-energy devices, if ever allowed to be mass distributed to consumers, coul d! Get your copy now before the Department of Energy bans this book!

292 PAGES. 8X10 PAPERBACK. ILLUSTRATED. BIBLIOGRAPHY. $16.95. COD H

THE ANTI-GRAVITY HANDBOOK

edited by David Hatcher Childress, with Nikola Tesla, T.B. Paulicki, Bruce Cathie, Albert Einstein and others

The new expanded compilation of material on Anti-Gravity, Free Energy, Flying Saucer Propulsion, UFOs, Suppressed Technology, NASA Cover-ups and more. Highly illustrated with patents, technical illustrations and photos. This revised and expanded edition has more material, including photos of Area 51, Nevada, the government's secret testing facility. This classic on weird science is back in a 90s format!

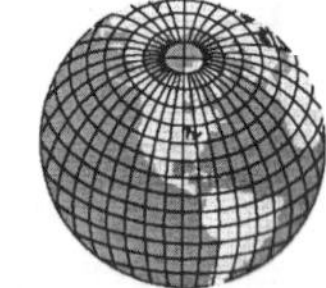

- **How to build a flying saucer.**
- **Arthur C. Clarke on Anti-Gravity.**
- **Crystals and their role in levitation.**
- **Secret government research and development.**
- **Nikola Tesla on how anti-gravity airships could draw power from the atmosphere.**
- **Bruce Cathie's Anti-Gravity Equation.**
- **NASA, the Moon and Anti-Gravity.**

230 PAGES. 7X10 PAPERBACK. BIBLIOGRAPHY/INDEX/APPENDIX. HIGHLY ILLUSTRATED. $14.95. CODE: AGH

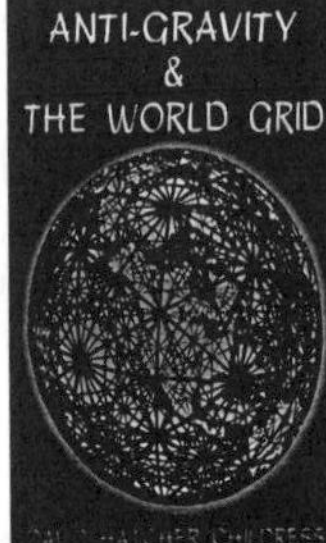

ANTI–GRAVITY & THE WORLD GRID

Is the earth surrounded by an intricate electromagnetic grid network offering free energy? This compilation of material on ley lines and world power points contains chapters on the geography, mathematics, and light harmonics of the earth grid. Learn the purpose of ley lines and ancient megalithic structures located on the grid. Discover how the grid made the Philadelphia Experiment possible. Explore the Coral Castle and many other mysteries, including acoustic levitation, Tesla Shields and scalar wave weaponry. Browse through the section on anti-gravity patents, and research resources.

274 PAGES. 7X10 PAPERBACK. ILLUSTRATED. $14.95. CODE: AGW

ANTI–GRAVITY & THE UNIFIED FIELD

edited by David Hatcher Childress

Is Einstein's Unified Field Theory the answer to all of our energy problems? Explored in this compilation of material is how gravity, electricity and magnetism manifest from a unified field around us. Why artificial gravity is possible; secrets of UFO propulsion; free energy; Nikola Tesla and anti-gravity airships of the 20s and 30s; flying saucers as superconducting whirls of plasma; anti-mass generators; vortex propulsion; suppressed technology; government cover-ups; gravitational pulse drive; spacecraft & more.

240 PAGES. 7X10 PAPERBACK. ILLUSTRATED. $14.95. CODE: AGU

ETHER TECHNOLOGY

A Rational Approach to Gravity Control

by Rho Sigma

This classic book on anti-gravity and free energy is back in print and back in stock. Written by a well-known American scientist under the pseudonym of "Rho Sigma," this book delves into international efforts at gravity control and discoid craft propulsion. Before the Quantum Field, there was "Ether." This small, but informative book has chapters on John Searle and "Searle discs;" T. Townsend Brown and his work on anti-gravity and ether-vortex turbines. Includes a forward by former NASA astronaut Edgar Mitchell.

108 PAGES. 6X9 PAPERBACK. ILLUSTRATED. $12.95. CODE: ETT

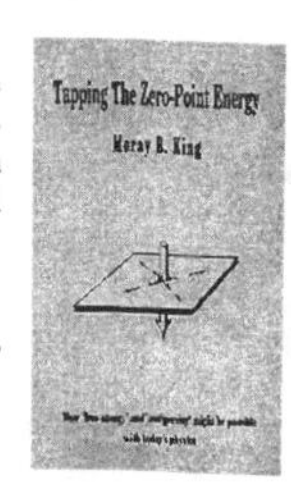

TAPPING THE ZERO POINT ENERGY

Free Energy & Anti-Gravity in Today's Physics

by Moray B. King

King explains how free energy and anti-gravity are possible. The theories of the zero point energy maintain there are tremendous fluctuations of electrical field energy imbedded within the fabric of space. This book tells how, in the 1930s, inventor T. Henry Moray could produce a fifty kilowatt "free energy" machine; how an electrified plasma vortex creates anti-gravity; how the Pons/Fleischmann "cold fusion" experiment could produce tremendous heat without fusion; and how certain experiments might produce a gravitational anomaly.

170 PAGES. 5X8 PAPERBACK. ILLUSTRATED. $9.95. CODE: TAP

24 hour credit card orders—call: 815-253-6390 fax: 815-253-6300

email: auphq@frontiernet.net www.adventuresunlimitedpress.com www.wexclub.com